797,885 Books
are available to read at

Forgotten Books

www.ForgottenBooks.com

Forgotten Books' App
Available for mobile, tablet & eReader

ISBN 978-1-330-02363-1
PIBN 10006549

This book is a reproduction of an important historical work. Forgotten Books uses
state-of-the-art technology to digitally reconstruct the work, preserving the original format
whilst repairing imperfections present in the aged copy. In rare cases, an imperfection in
the original, such as a blemish or missing page, may be replicated in our edition. We do,
however, repair the vast majority of imperfections successfully; any imperfections that
remain are intentionally left to preserve the state of such historical works.

Forgotten Books is a registered trademark of FB &c Ltd.
Copyright © 2015 FB &c Ltd.
FB &c Ltd, Dalton House, 60 Windsor Avenue, London, SW19 2RR.
Company number 08720141. Registered in England and Wales.

For support please visit www.forgottenbooks.com

1 MONTH OF FREE READING

at

www.ForgottenBooks.com

By purchasing this book you are eligible for one month membership to ForgottenBooks.com, giving you unlimited access to our entire collection of over 700,000 titles via our web site and mobile apps.

To claim your free month visit: www.forgottenbooks.com/free6549

* Offer is valid for 45 days from date of purchase. Terms and conditions apply.

Similar Books Are Available from
www.forgottenbooks.com

General Mathematics
by Raleigh Schorling

Algebra for Beginners
With Numerous Examples, by Isaac Todhunter

The Principles of Mathematics
by Bertrand Russell

Mathematical Exercises for Home Work
by A. T. Richardson

A Treatise on Algebra
by Charles Smith

Elements of the Method of Least Squares
by Mansfield Merriman

Higher Algebra
by George Egbert Fisher

Linear Algebra
by Hussein Tevfik

Lectures on Fundamental Concepts of Algebra and Geometry
by John Wesley Young

Mathematical Tracts, Vol. 1
by Francis William Newman

Durell's Algebra, Vol. 1 of 2
Two Book Course, by Fletcher Durell

Diophantos of Alexandria
A Study in the History of Greek Algebra, by Thomas Little Heath

A Complete Course in Algebra
For Academies and High Schools, by Webster Wells

On the Study and Difficulties of Mathematics
by Augustus de Morgan

The Constructive Development of Group-Theory
With a Bibliography, by Burton Scott Easton

College Algebra
With Applications, by E. J. Wilczynski

An Introduction to Algebra
Being the First Part of a Course of Mathematics, by Jeremiah Day

On Mathematics and Mathematicians
by Robert Edoward Moritz

Exercises in Algebra
by Edward R. Robbins

The High School Course in Mathematics
by Ernest B. Skinner

Educ T 129.00.397

Harvard College Library
THE GIFT OF
GINN AND COMPANY

RUDIMENTS OF ALGEBRA

BY

GEORGE EGBERT FISHER, M.A., Ph.D.

AND

ISAAC J. SCHWATT, Ph.D.

ASSISTANT PROFESSORS OF MATHEMATICS IN THE
UNIVERSITY OF PENNSYLVANIA

PHILADELPHIA
FISHER AND SCHWATT
1900

Educ T 29.00.397

GIFT OF
GINN & COMPANY
MARCH 17, 19??

COPYRIGHT, 1900,
BY FISHER AND SCHWATT.

PREFACE.

This book gives a brief course in the elementary processes of algebra. Great care has been given to the representation, and to the solutions of typical exercises in the text, inculcating a thorough knowledge of algebraic processes.

Each principle or method is first clearly illustrated by numerous simple examples. But it is nowhere assumed that the principles are thereby proved. Even a beginner should not be encouraged, by text-book or teacher, to accept an illustrative example as a proof, or he will lose much of the educational value of the study.

Nearly all of the exercises have been prepared expressly for this book. They have been carefully graded, and, it is believed, none of them are too difficult for the average beginner.

The introductory chapter extends the familiar processes of arithmetic to the corresponding processes of algebra. The pupil is led by simple exercises, similar to those in arithmetic, to understand the use of letters to represent general and unknown numbers. Negative numbers are naturally introduced in connection with the extension of subtraction of arithmetical numbers. The meaning and use of positive and negative numbers, in the fundamental operations, are properly emphasized.

Equations and problems are distributed throughout the book. The importance of equivalent equations is not overlooked, but is very briefly and simply considered in Chapter IV. Until that chapter is reached, the solutions of equations should be checked.

All the matter in the book is printed in large type, and much pains have been taken to make the pages open and attractive.

Any suggestions from teachers and others will be much appreciated.

G. E. F.
L. J. S.

UNIVERSITY OF PENNSYLVANIA,
PHILADELPHIA.

CONTENTS.

CHAPTER I.

	PAGE
INTRODUCTION	1
General Number	1
Equations without Transposition	8
Problems	10
Positive and Negative Numbers	18

CHAPTER II.

FUNDAMENTAL OPERATIONS WITH ALGEBRAIC NUMBERS	23
Addition	23
Subtraction	24
Multiplication	26
Division	28
Parentheses	32
Positive Integral Powers	35

CHAPTER III.

FUNDAMENTAL OPERATIONS WITH INTEGRAL ALGEBRAIC EXPRESSIONS	38
Addition and Subtraction	39
Parentheses	46
Equations with Transposition, and Problems	47
Multiplication	51
Equations and Problems	57
Division	61

CHAPTER IV.

INTEGRAL ALGEBRAIC EQUATIONS	68
Equivalent Equations	70
Problems	72

CHAPTER V.

Type-Forms in Multiplication and Division 76

CHAPTER VI.

Factors and Multiples of Integral Algebraic Expressions 82
    Integral Algebraic Factors 82
    Highest Common Factors 91
    Lowest Common Multiples 93
    Solution of Equations by Factoring 95

CHAPTER VII.

Fractions 97
    Reduction to Lowest Terms 98
    Reduction to Lowest Common Denominator . . . 99
    Equations and Problems 102
    Addition and Subtraction 106
    Multiplication 111
    Division 114
    Complex Fractions 116

CHAPTER VIII.

Fractional Equations 118
    Problems 120

CHAPTER IX.

Literal Equations 124
    General Problems 125

CHAPTER X.

Simultaneous Linear Equations 128
    Elimination by Addition and Subtraction . . . 129
    Elimination by Comparison 132
    Elimination by Substitution 133
    Linear Equations in Three Unknown Numbers . 135
    Problems 136

CHAPTER XI.

	PAGE
INVOLUTION	141
Powers of Monomials	142
Powers of Binomials	143

CHAPTER XII.

EVOLUTION	144
Roots of Monomials	147
Square Roots of Multinomials	148
Cube Roots of Multinomials	150
Roots of Arithmetical Numbers	153
Irrational Roots	159

CHAPTER XIII.

QUADRATIC EQUATIONS	160
Pure Quadratic Equations	160
Solution by Factoring	162
Solution by Completing the Square	163
Problems	164

CHAPTER XIV.

SIMULTANEOUS QUADRATIC EQUATIONS	168
Problems	171

CHAPTER XV.

PROGRESSIONS	173
Arithmetical Progression	173
Geometrical Progression	179

5. As was assumed in Art. 2, the operations of Addition, Subtraction, Multiplication, and Division are denoted by the same symbols in Algebra as in Arithmetic.

Just as $5 + 3$, read *five plus three*, means that 3 is to be added to 5; so $a + b$, read *a plus b*, means that b is to be added to a.

Just as $5 - 3$, read *five minus three*, means that 3 is to be subtracted from 5; so $a - b$, read *a minus b*, means that b is to be subtracted from a.

Just as 5×3, read *five multiplied by three*, means that 5 is to be multiplied by 3; so $a \times b$, read *a multiplied by b*, means that a is to be multiplied by b.

A dot (·) is frequently used, instead of the symbol $\times$, to denote multiplication; as $a \cdot b$ for $a \times b$.

Just as $10 \div 5$, read *ten divided by five*, means that 10 is to be divided by 5; so $a \div b$, read *a divided by b*, means that a is to be divided by b.

6. The symbol of multiplication between two literal numbers, or one literal number and an Arabic numeral, is frequently omitted.

E.g., the product $x \times y \times z$, or $x \cdot y \cdot z$, is usually written, xyz, and is read *x-y-z*.

But the symbol of multiplication between two numerals cannot be omitted without changing the meaning.

E.g., if in the indicated multiplication, 3×6, or $3 \cdot 6$, the symbol, $\times$, or ·, were omitted, we should have 36, not 18.

7. In a chain of additions and subtractions the operations are to be performed successively from left to right.

E.g., $\quad 7 + 4 - 3 + 2 = 11 - 3 + 2 = 8 + 2 = 10.$

In a chain of multiplications and divisions the operations are to be performed successively from left to right.

E.g., $\quad 12 \times 2 \div 3 \times 4 = 24 \div 3 \times 4 = 8 \times 4 = 32.$

8. An **Algebraic Expression** is a number expressed by means of the signs and symbols of Algebra; as x, mn, $ab - cd$, etc.

9. The **Symbol of Equality**, $=$, read *is equal to*, is placed between two numbers to indicate that they have the same or equal values; as $3 + 2 = 5$.

10. The **Symbol of Inequality**, $>$, read *is greater than*, is used to indicate that the number on its left is greater than that on its right; as $7 > 5$.

11. The **Symbol of Inequality**, $<$, read *is less than*, is used to indicate that the number on its left is less than that on its right; as $3 < 4 + 2$.

12. The use of letters to represent general numbers may be illustrated by a few simple examples.

Ex. **1.** If a boy has 3 books and is given 2 more, he will have $3 + 2$ books. If he has a books and is given 5 more, he will have $a + 5$ books. If he has m books and is given n more, he will have $m + n$ books.

Ex. **2.** If a man buys 5 city lots at 120 dollars each, he pays 120×5 dollars for the lots. If he buys a lots at 150 dollars each, he pays $150\,a$ dollars for the lots. If he buys u lots at v dollars each, he pays vu dollars for the lots.

Ex. **3.** If a train runs 60 miles in 2 hours, it runs $60 \div 2$ miles in 1 hour. If it runs a miles in 5 hours, it runs $a \div 5$ miles in 1 hour. If it runs p miles in q hours, it runs $p \div q$ miles in 1 hour.

Ex. **4.** If, in a number of *two* digits, the digit in the *units'* place is 3 and the digit in the *tens'* place is 5, the number is $10 \times 5 + 3$. If the digit in the units' place is a and the digit in the tens' place is b, the number is $10\,b + a$.

Ex. **5.** Just as $2 = 1 + 1$, and $3 = 1 + 1 + 1$,

so $\qquad 2\,a = a + a$, and $3\,a = a + a + a$.

Therefore, just as $3 + 2 = 5$, so $3\,a + 2\,a = 5\,a$.

In like manner, $\qquad 5\,x - 3\,x = 2\,x$;

and $\qquad \tfrac{1}{2} x + \tfrac{2}{3} x = \tfrac{7}{6} x$.

EXERCISES II.

Read each of the following expressions:

1. $x+y$.
2. $a-b$.
3. $b \times c$.
4. $y \div z$.
5. $2x+3y$.
6. $4x-9y$.
7. $3a \times 2b$.
8. $9m \div 3n$.
9. $a+b+c$.
10. $a+b-c$.
11. $a-b-c$.

12. The width of a room is a feet, and the length is b feet more than the width. What is the length of the room?

13. A man is now x years old. How old will he be in 10 years? In y years?

14. A piece of cloth contains a yards. If 15 yards are cut off, how many yards are left? If c yards are cut off, how many yards are left?

15. A man is now n years old. How old was he 15 years ago? m years ago? How long must he live to be 90 years old? To be a years old?

16. If 1 pound of tea costs 75 cents, how much do 3 pounds cost? d pounds?

17. If 1 pound of tea costs c cents, how much do 5 pounds cost? x pounds?

18. A boy paid 20 cents for 5 pencils. How much did he pay for each?

19. A boy paid m cents for 6 pencils. How much did he pay for each? If he paid m cents for x pencils, how much did he pay for each?

20. 10×2, 10×3, 10×4, etc., are particular multiples of 10. Express *any* multiple of 10.

21. Write a number containing 4 units and 7 tens. Containing x units and y tens.

22. Write a number containing 3 units, 5 tens, and 9 hundreds. Containing a units, b tens, and c hundreds.

23. One boy has $5x$ apples and another has $3x$ apples. How many apples have both boys? How many apples has one more than the other?

24. A man paid $15a$ dollars for a house and $6a$ dollars for a lot. How much did he pay for both? How much more for the house than for the lot?

What are the values of the following expressions:

25. $x + x.$
26. $a + 2a.$
27. $z + 3z.$
28. $4d + 3d.$
29. $8c + 9c.$
30. $7y + 3y.$
31. $n + \frac{1}{3}n.$
32. $\frac{1}{2}m + \frac{2}{3}m.$
33. $x + 2x + 6x.$
34. $y + 2y + 3y.$
35. $4z + 9z + 6z.$
36. $x - x.$
37. $2a - a.$
38. $5y - 2y.$
39. $\frac{3}{8}b - \frac{1}{2}b.$
40. $3n + 9n - 6n.$
41. $5z - 2z + 6z.$
42. $9c - 4c - 2c.$

Axioms.

13. An **Axiom** is a truth so simple that it cannot be made to depend upon a truth still simpler.

Algebra makes use of the following mathematical axioms:

(i.) *Every number is equal to itself.* E.g., $7 = 7$, $a = a$.

(ii.) *The whole is equal to the sum of all its parts.*

E.g., $5 = 1 + 1 + 1 + 1 + 1,\ 7 = 3 + 4.$

(iii.) *If two numbers be equal, either can replace the other in any algebraic expression in which it occurs.*

E.g., if $a + b = c$, and $b = 2$, then $a + 2 = c$, replacing b by 2.

(iv.) *Two numbers which are each equal to a third number are equal to each other.*

E.g., if $a = b$, and $c = b$, then $a = c$.

(v.) *The whole is greater than any of its parts; and, conversely, any part is less than the whole.*

E.g., $3 + 2 > 2$ and $2 < 3 + 2$.

14. *Literal numbers*, as has been stated, are numbers which may have *any values whatever*. But it is frequently necessary to assign particular values to such numbers.

15. Substitution is the process of replacing a literal number in an algebraic expression by a particular value. See axiom (iii.). Simple examples in substitution have already been given in Art. 2.

Ex. **1.** If, in $a + b$, we let $a = 3$ and $b = 5$, then
$$a + b = 3 + 5 = 8, \text{ or } a + b = 8.$$

Ex. **2.** If, in $a + b - 2a + 3b - c$, we let $a = 6$, $b = 11$, $c = 1$, we have
$$a + b - 2a + 3b - c = 6 + 11 - 2 \times 6 + 3 \times 11 - 1$$
$$= 6 + 11 - 12 + 33 - 1 = 37.$$

Ex. **3.** If, in the last example, $a = 3$, $b = 1$, and $c = 1$, we have $a + b - 2a + 3b - c = 3 + 1 - 6 + 3 - 1 = 4 - 6 + 3 - 1$.

We cannot further reduce $4 - 6 + 3 - 1$, since we are unable, *as yet*, to subtract 6 from 4.

EXERCISES III.

When $a = 6$, $b = 3$, $c = 2$, find the values of the following expressions:

1. $a + b$.
2. $a - b$.
3. ab.
4. $a + b$.
5. $a - b + c$.
6. $a - b - c$.
7. abc.
8. $a + b \times c$.
9. $5a$.
10. $4ac$.
11. $2aa$.
12. $3bbb$.
13. $c + 3a$.
14. $4b - 2c$.
15. $2ab + 5ac$.
16. $6ac - 3ab$.
17. $2a + 3b - 4c$.
18. $ab + bc - ac$.
19. $ab - ac - bc$.
20. $5ab - 3ac + 7bc$.

Fundamental Principles.

16. The following principles are obtained directly from the axioms:

(i.) *If the same number, or equal numbers, be added to equal numbers, the sums will be equal.*

(ii.) *If the same number, or equal numbers, be subtracted from equal numbers, the remainders will be equal.*

(iii.) *If equal numbers be multiplied by the same number, or by equal numbers, the products will be equal.*

(iv.) *If equal numbers be divided by the same number (except 0), or by equal numbers, the quotients will be equal.*

E.g., if $3x = 6$,

then $3x + 2 = 6 + 2$, $3x - 5 = 6 - 5$,

$3x \times 4 = 6 \times 4$, $3x \div 3 = 6 \div 3$.

Equations.

17. An **Equation** is a statement that two expressions are equal; as $7 \times 9 = 63$, $4 \times 7 + 3 = 31$.

The **First Member** of an equation is the expression on the *left* of the symbol $=$; the **Second Member** is the expression on the *right* of the symbol $=$.

18. Ex. 1. What is the value of x in the equation

$$3x + 8x = 22?$$

Since $3x + 8x = 11x$, we have

$$11x = 22.$$

Dividing both members by 11, [Art. 16, (iv.)],

$$x = 2.$$

To check this result we substitute 2 for x in the equation,

$$3x + 8x = 3 \times 2 + 8 \times 2 = 6 + 16 = 22.$$

Ex. 2. If $8x - 3x$ has the value 20, what is the value of x?

We have $\qquad 8x - 3x = 20.$

Or, since $8x - 3x = 5x,\qquad 5x = 20.$

Dividing both members by 5, $x = 4.$

Check: $8 \times 4 - 3 \times 4 = 32 - 12 = 20.$

19. An **Unknown Number** of an equation is a number whose value is to be found from the equation.

The **Known Numbers** of an equation are the numbers whose values are given.

In the equation $\qquad x + 1 = 3$

the unknown number is x, and the known numbers are 1 and 3.

Unknown numbers are usually represented by the final letters of the alphabet, x, y, z, etc., as in the above examples.

EXERCISES IV.

Find the value of x in each of the following equations:

1. $2x = 8.$
2. $5x = 15.$
3. $3x = 0.$
4. $\frac{1}{2}x = 3.$
5. $\frac{1}{3}x = 7.$
6. $\frac{1}{4}x = 0.$
7. $\frac{2}{3}x = 8.$
8. $\frac{3}{2}x = 9.$
9. $x + 2x = 6.$
10. $2x - x = 5.$
11. $x + 7x = 24.$
12. $7x - x = 18.$
13. $2x + 3x = 10.$
14. $4x - 2x = 8.$
15. $x + 2x + 3x = 30.$
16. $3x - 2x + x = 2.$
17. $4x + 3x + 5x = 24.$
18. $6x + 2x - 3x = 20.$
19. $7x - 3x + 5x = 81.$
20. $9x - 4x - 2x = 21.$
21. $x + \frac{1}{3}x = 4.$
22. $x - \frac{1}{3}x = 4.$
23. $x + \frac{3}{4}x = 14.$
24. $x - \frac{2}{5}x = 6.$
25. $2x + \frac{3}{2}x = 14.$
26. $\frac{10}{3}x - 3x = 2.$
27. $x + \frac{1}{2}x + \frac{1}{3}x = 11.$
28. $x - \frac{1}{2}x + \frac{1}{3}x = 10.$
29. $2x + \frac{1}{3}x + \frac{3}{4}x = 7\frac{1}{4}.$
30. $3x - \frac{2}{3}x - \frac{3}{2}x = 5.$

Problems solved by Equations.

20. A Problem is a question proposed for solution.

Another use of literal numbers is shown by the following problems:

Pr. 1. The older of two brothers has twice as many marbles as the younger, and together they have 33 marbles. How many has the younger?

The number of marbles the younger brother has is, as yet, an *unknown number*.

Let us represent this unknown number by some letter, say x.

Then, since the older brother has twice as many, he has $2x$ marbles.

The problem states,

in *verbal* language: *the number of marbles the younger has plus the number the older has is equal to 33*;

in *algebraic* language, $x + 2x = 33$,

or, $\quad\quad\quad\quad\quad 3x = 33$.

Dividing both members of the last equation by 3, we have

$$x = 11,$$

the number of marbles the younger has.

The older has, $2x, = 2 \times 11, = 22$ marbles.

To check this result, we substitute 11 for x in the equation of the problem:
$$x + 2x = 11 + 22 = 33.$$

Notice that the letter x stands for an abstract number. The beginner must never put x for marbles, distance, time, etc., but for the *number* of marbles, of miles, of hours, etc.

Pr. 2. Divide 52 into three parts, so that the second shall be one-half of the first, and the third one-fourth of the second.

GENERAL NUMBERS.

Let x stand for the first part.

Then $\frac{1}{2}x$ stands for the second part,

and $\frac{1}{4} \times \frac{1}{2}x, = \frac{1}{8}x$, stands for the third part.

The problem states,

in *verbal* language: *the first part, plus the second part, plus the third part, is equal to* 52;

in *algebraic* language: $x + \frac{1}{2}x + \frac{1}{8}x = 52$,

or, $\frac{13}{8}x = 52.$

Dividing both members of the last equation by 13,

$$\frac{1}{8}x = 4.$$

Multiplying both members of this equation by 8,

$x = 32$, the first part.

Then the second part is

$$\frac{1}{2}x, = \frac{1}{2} \times 32, = 16,$$

and the third part is

$$\frac{1}{8}x, = \frac{1}{8} \times 32, = 4.$$

Check: $x + \frac{1}{2}x + \frac{1}{8}x = 32 + 16 + 4 = 52.$

21. In stating problems in algebraic language, the beginner should observe the following directions:

(i.) *Read the problem carefully, and note what are the numbers whose values are required.*

(ii.) *Let some letter, say x, stand for one of the required numbers.*

(iii.) *The problem will contain statements about the values of other numbers. Use these statements to express their values in terms of x.*

(iv.) *Express concisely in verbal language a statement in the problem which furnishes an equation.*

(v.) *Express this statement in algebraic language.*

EXERCISES V.

1. What number is seven times x? Eleven times x?

2. Seven times a number is 126. What is the number?

3. Eleven times a number is 110. What is the number?

4. The greater of two numbers is five times the less. If the less is x, what is the greater? What is their sum? Their difference?

5. The greater of two numbers is five times the less. If their sum is 84, what are the numbers?

6. The greater of two numbers is three times the less. If their difference is 26, what are the numbers?

7. A father is twice as old as his son. If the son is x years old, how old is the father? What is the sum of their ages? How much older is the father than the son?

8. A father is six times as old as his son, and the sum of their ages is 35 years. How old is each?

9. A father is three times as old as his son. If the father is 30 years older than his son, what are their ages?

10. At an election A received twice as many votes as B, and his majority was 450. How many votes did each receive?

11. In a company are 25 persons. The number of children is four times the number of adults. How many are there of each?

12. Two trains leave Philadelphia in opposite directions. After one hour they are 69 miles apart. If one has gone twice as far as the other, how many miles is each from Philadelphia?

13. Two trains leave Chicago in the same direction. After one hour they are 36 miles apart. If one has gone three times as far as the other, how far is each from Chicago?

14. A man pays $77 in one-dollar bills and ten-dollar bills. If he pays the same number of one-dollar bills as of ten-dollar bills, how many of each does he pay?

15. In a number of two digits, the tens' digit is twice the units' digit, and their sum is 12. What are the digits? What is the number?

16. In a number of two digits, the units' digit is three times the tens' digit, and their difference is 4. What is the number?

17. What is the sum of twice x and five times x? The difference?

18. If twice a number is added to five times the same number, the sum will be 56. What is the number?

19. If three times a number is subtracted from eight times the same number, the remainder will be 45. What is the number?

20. A man paid $72 for equal quantities of large and small coal. He paid $5 a ton for the large, and $3 a ton for the small. How many tons of each did he buy?

21. A man bought equal quantities of large and small coal. He paid $5 a ton for the large and $3 a ton for the small. If he paid $18 more for the large than for the small, how many tons of each did he buy?

22. A traveller first rides his bicycle 8 miles an hour. He then rides the same number of hours in a car 30 miles an hour. If he travels 114 miles, how many hours did he ride his bicycle?

23. Two trains run out of New York in opposite directions. One runs 40 miles an hour, the other 35 miles an hour. After how many hours will they be 300 miles apart?

24. Two trains run out of New York in the same direction. One runs 45 miles an hour, the other 32 miles an hour. After how many hours will they be 65 miles apart?

25. A boy has 90 cents in dimes and five-cent pieces. He has the same number of dimes as of five-cent pieces. How many coins of each kind has he?

26. A owes B $32. He pays his debt in ten-dollar bills, and receives in change the same number of two-dollar bills. How many ten-dollar bills did A pay B?

27. A cistern has two pipes. One lets in 10 gallons a minute, and the other 14 gallons a minute. If the cistern holds 216 gallons, how many minutes will it take the pipes to fill it?

28. A cistern has two pipes. One lets in 12 gallons a minute, and the other lets out 8 gallons a minute. How many minutes will it take the one pipe to let in 68 gallons more than the other lets out?

29. What is the sum of x, three times x, and five times x? Of x, twice x, and seven times x?

30. The sum of a certain number, three times the number, and five times the number is 99. What is the number?

31. Three boys, A, B, and C, together have 12 pencils. B has twice as many as A, and C three times as many as A. How many has A? How many has each?

32. Divide 189 into three parts, so that the second part shall be three times the first, and the third part five times the first.

33. A merchant receives $96 in ten-dollar bills, five-dollar bills, and one-dollar bills. He receives the same number of each kind. How many of each does he receive?

34. At an election 666 votes were cast. A, B, and C were candidates. B received twice as many votes as C, and A three times as many as C. How many votes did each receive?

35. A cistern has 3 pipes. The first lets in 5 gallons a minute, the second 7 gallons a minute, and the third 10 gallons a minute. If the cistern holds 242 gallons, how long will it take the pipes to fill it?

36. A cistern has 3 pipes. The first lets in 6 gallons a minute, the second 12 gallons a minute, and the third lets out 8 gallons a minute. How many minutes will it take the two pipes to let in 120 gallons more than the third pipe lets out?

37. An estate of $7200 is divided among 2 sons and 2 daughters. The sons receive equal amounts, and a daughter receives twice as much as a son. How many dollars does each receive?

38. A man pays $90 in ten-dollar bills, five-dollar bills, two-dollar bills, and one-dollar bills. He pays the same number of each kind. How many of each does he pay?

39. What is twice $3x$? Seven times $5x$? Four times $9x$?

40. A receives x dollars, B receives twice as much as A, and C receives 4 times as much as B. How many dollars does C receive? How many dollars do all receive?

41. Three boys, A, B, and C, together receive $55. B receives twice as much as A, and C four times as much as B. How many dollars does each receive?

42. A merchant's profits doubled each year for three years. If his profits for the three years were $5250, what were his profits the first year?

43. In a company are 36 persons. The number of women is twice the number of men, and the number of children is three times the number of women. How many of each are in the company?

44. What number is $\frac{1}{3}$ of x? $\frac{3}{4}$ of x?

45. If $\frac{1}{3}$ of a number is 18, what is the number?

46. The less of two numbers is $\frac{2}{5}$ of the greater. If the greater is x, what is the less? What is their sum? Their difference?

47. The less of two numbers is $\frac{2}{5}$ of the greater. If their sum is 49, what are the numbers?

48. A and B together have $800. If B has $\frac{3}{5}$ as much as A, how many dollars has each?

49. A has $50 more than B. If B has $\frac{2}{3}$ as much as A, how many dollars has each?

50. Two boys, A and B, catch 24 fish. If A catches $\frac{5}{3}$ as many as B, how many fish does each catch?

51. A workman pays $\frac{2}{3}$ of his wages for board. If he has left $6 each week, what are his wages?

52. Two boys together solve 64 problems. If the first solves $\frac{7}{9}$ as many as the second, how many problems does each solve?

53. A solves 28 more problems than B. If B solves $\frac{3}{7}$ as many as A, how many problems does each solve?

54. A tree 88 feet high is broken by the wind. If the part left standing is $\frac{2}{9}$ of the part broken off, how long is each part?

55. What is the sum of $\frac{1}{4}$ of x and $\frac{2}{3}$ of x? The difference?

56. If $\frac{1}{4}$ of a number is added to $\frac{2}{3}$ of the same number, the sum will be 44. What is the number?

57. If $\frac{2}{3}$ of a number is subtracted from $\frac{3}{4}$ of the same number, the remainder will be 5. What is the number?

58. If to a number is added $\frac{1}{2}$ of itself and $\frac{2}{3}$ of itself, the sum will be 78. What is the number?

59. Three boys, A, B, and C, together have 51 pencils. B has $\frac{3}{4}$ as many as A, and C has $\frac{4}{5}$ as many as A. How many pencils has each?

60. Divide 105 into 3 parts, so that the first shall be twice the second, and the third $\frac{1}{2}$ of the second.

61. A man makes a journey of 171 miles. He goes $\frac{4}{5}$ as far by boat as by train, and $\frac{1}{10}$ as far by stage as by train. How many miles does he go by each conveyance?

62. What is $\frac{1}{8}$ of four times x? Twice $\frac{3}{5}$ of x? $\frac{2}{3}$ of $\frac{4}{5}$ of x?

63. The second of three numbers is four times the first, and the third is $\frac{1}{3}$ of the second. If the first number is x, what is the second? The third? What is the sum of the three numbers?

64. The sum of three numbers is 63. The second is twice the first, and the third is $\frac{3}{5}$ of the second. What are the numbers?

65. A library contains 4700 volumes. The number of volumes of history is ⅛ of the number of volumes of science, and the number of volumes of fiction is $\frac{3}{11}$ of the number of volumes of history. How many volumes of each kind are in the library?

66. The width of a field is ⅔ of its length, and the distance around it is 96 rods. What is the width and the length of the field?

67. The sum of $480 is divided among A, B, and C. B receives ⅔ as much as A, and C as much as A and B together. How many dollars does each receive?

68. A sells a number of apples at 3 cents apiece, and B ¾ as many at 2 cents apiece. If they receive together 88 cents, how many apples does each sell?

Parentheses.

22. Parentheses, ( ), and **Brackets**, [ ], are used to indicate that whatever is placed within them is to be treated as a whole.

E.g., $10 - (2 + 5)$ means that the result of adding 5 to 2, or 7, is to be subtracted from 10; that is,

$$10 - (2 + 5) = 10 - 7 = 3.$$

But $10 - 2 + 5$ means that 2 is to be subtracted from 10 and 5 is then to be added to that result; that is,

$$10 - 2 + 5 = 8 + 5 = 13.$$

In like manner, $[27 - (3 + 2) \times 5] \div 2$ means that the result of multiplying the sum $3 + 2$ by 5 is first to be subtracted from 27, and the remainder is then to be divided by 2; that is,

$$[27 - (3 + 2) \times 5] \div 2 = [27 - 25] \div 2 = 2 \div 2 = 1.$$

Likewise, the result of multiplying $a + b$ by c is $(a + b)c$, etc.

EXERCISES VI.

Find the values of the following expressions:

1. $9+(5+2)$.
2. $9-(5+2)$.
3. $9+(5-2)$.
4. $9-(5-2)$.
5. $7+(4+3)$.
6. $7+(4-3)$.
7. $7-(4+3)$.
8. $7-(4-3)$.
9. $3(8+6)$.
10. $3(8-6)$.
11. $(8+6) \div 2$.
12. $(8-6) \div 2$.
13. $(16+8)+(9+3)$.
14. $(16+8)-(9+3)$.
15. $(16-8)+(9-3)$.
16. $(16-8)-(9-3)$.
17. $12-[9+(4-3)]$.
18. $16-[(8-3)-(4-1)]$.
19. $[24+(11-5)] \times 2$.
20. $[21-(13-7)] \div 3$.

When $a=8$, $b=4$, $c=2$, find the values of:

21. $a+(b-c)$.
22. $a-(b+c)$.
23. $a-(b-c)$.
24. $c+3(a-b)$.
25. $b[a-(b-c)]$.
26. $a-2(b-c)$.
27. $a[a-(b+c)]$.
28. $[a-(b-c)] \div c$.
29. $[b+(a-c)] \div c$.

POSITIVE AND NEGATIVE NUMBERS, OR ALGEBRAIC NUMBERS.

23. In ordinary Arithmetic we subtract a number from an equal or a greater number. We are familiar with such operations as

5	4	3	minuend
3	3	3	subtrahend	(i.)
2	1	0	remainder

But such operations as

2	1	0	minuend
3	3	3	subtrahend	(ii)
?	?	?	remainder

have not occurred in ordinary Arithmetic. In Arithmetic we cannot subtract from a number more units than are contained in the number.

24. Now, in the operations (i.) above, the remainder 2 indicates that the subtrahend is *two* units *less* than the minuend; the remainder 1 that the subtrahend is *one* unit *less* than the minuend; and the remainder 0 that the subtrahend is *equal* to the minuend.

In the operations (ii.), we must indicate by the remainders that the subtrahend is *one, two, three*, etc., units *greater* than the corresponding minuend.

We do this by placing the sign ⁻ before the symbols for *one, two, three*, etc.; as ⁻1, ⁻2, ⁻3, etc.

The remainders in these cases are called **Negative Numbers**; as ⁻1, ⁻2, ⁻3, etc., read *negative one, negative two, negative three*, etc.

For the sake of distinction, the remainders in the operations (i.) are called **Positive Numbers**.

They are indicated by the sign ⁺; as ⁺1, ⁺2, ⁺3, etc., read *positive one, positive two, positive three*, etc.

Positive and negative numbers are called **Algebraic** or **Relative Numbers**.

25. It is important to notice that a negative remainder does not mean that more units have been taken from the minuend than were contained in it; *such a remainder indicates that the subtrahend is greater than the minuend by as many units as are contained in the remainder.*

26. We can now write (i.) and (ii.) as follows:

pos. 5	pos. 4	pos. 3	pos. 2	pos. 1	0	min.
pos. 3	pos. 3	pos. 3	pos. 3	pos. 3	pos. 3	sub.
pos. 2	pos. 1	0	neg. 1	neg. 2	neg. 3	rem.

or

+5	+4	+3	+2	+1	0	min.
+3	+3	+3	+3	+3	+3	sub.
+2	+1	0	-1	-2	-3	rem.

(iii)

27. We thus have in Algebra the series of numbers,

$$\ldots, {}^+5, {}^+4, {}^+3, {}^+2, {}^+1, 0, {}^-1, {}^-2, {}^-3, {}^-4, {}^-5, \ldots,$$

wherein the signs, ·, indicate that the succession of numbers continues without end in both directions. This series is usually written with the positive numbers on the right, as

$$\ldots, {}^-5, {}^-4, {}^-3, {}^-2, {}^-1, 0, {}^+1, {}^+2, {}^+3, {}^+4, {}^+5, \ldots.$$

28. In this series the numbers increase by *one* from left to right, and decrease by *one* from right to left. Or, a number is greater than any number on its left, and less than any number on its right.

Thus, $^+2$ is one unit greater than $^+1$, two units greater than 0, three units greater than $^-1$, etc. Again, $^-3$ is three units greater than $^-6$, two units less than $^-1$, three units less than 0, etc.

29. The signs $^+$ and $^-$ are called signs of *quality;* the signs $+$ and $-$, signs of *operation*. The two sets of signs must, as yet, be carefully distinguished.

30. The **Absolute Value** of a number is the number of units contained in it without regard to their *quality*.

E.g., the absolute value of $^+4$ is 4, of $^-5$ is 5.

31. Negative numbers have been introduced by extending the operation of subtraction. But it is necessary to treat them as numbers apart from this particular operation.

As in Arithmetic, so in Algebra, any integer is an aggregate of like units.

Just as $\qquad 4 = 1 + 1 + 1 + 1,$

so $\quad {}^+4 = {}^+1 + {}^+1 + {}^+1 + {}^+1,$ and $^-4 = {}^-1 + {}^-1 + {}^-1 + {}^-1.$

Just as $\qquad \frac{2}{3} = \frac{1}{3} + \frac{1}{3},$

so $\quad {}^+(\frac{2}{3}) = {}^+(\frac{1}{3}) + {}^+(\frac{1}{3}),$ and $^-(\frac{2}{3}) = {}^-(\frac{1}{3}) + {}^-(\frac{1}{3}).$

EXERCISES VII.

Simplify the following expressions:

1. $^+5 - {}^+3.$
2. $^+5 - {}^+5.$
3. $^+5 - {}^+7.$
4. $^+11 - {}^+6.$
5. $^+11 - {}^+11.$
6. $^+11 - {}^+16.$

What value of x will make the first member of each of the following equations the same as the second:

7. $x - {}^+4 = {}^+3$. **8.** $x - {}^+4 = 0$. **9.** $x - {}^+4 = {}^-4$.

How many units is each of the following numbers greater or less than 0:

10. ${}^+4$. **11.** ${}^-3$. **12.** ${}^+11$. **13.** ${}^-9$. **14.** ${}^-11$.

Positive and Negative Numbers are Opposite Numbers.

32. In Arithmetic we have: *the remainder added to the subtrahend is equal to the minuend.* This principle, like all principles of Arithmetic, is retained in Algebra. We therefore have from (iii.) Art. 26·

${}^+3$	${}^+3$	${}^+3$	${}^+3$	${}^+3$	${}^+3$	sub.
${}^+2$	${}^+1$	0	${}^-1$	${}^-2$	${}^-3$	rem.
${}^+5$	${}^+4$	${}^+3$	${}^+2$	${}^+1$	0	min.

33. The equation ${}^+3 + {}^-3 = 0$ gives us the following important principle·

The sum of a positive number and a negative number having the same absolute value is equal to zero; i.e., *two such numbers cancel each other when united by addition.*

E.g., ${}^+1 + {}^-1 = 0$, ${}^+3 + {}^-3 = 0$, ${}^-17\frac{1}{2} + {}^+17\frac{1}{2} = 0$.

In general, ${}^+n + {}^-n = 0$.

For this reason, positive and negative numbers in their relation to each other are called *opposite* numbers. When their absolute values are equal, they are called *equal* and *opposite* numbers.

34. Any *quantities* which in their relation to each other are *opposite,* may be represented in Algebra by *positive* and *negative* numbers; as *credits* and *debits, gain* and *loss.*

Ex. 1. 100 *dollars credit* and 100 *dollars debit* cancel each other. That is, 100 *dollars credit united with* 100 *dollars debit is equal to neither credit nor debit;* or,

100 *dollars credit* + 100 *dollars debit* = *neither credit nor debit.*

If *credits* be taken *positively* and *debits negatively*, then 100 *dollars credit* may be represented by $^+100$, and 100 *dollars debit* by $^-100$. Their united effect, as stated above, may then be represented algebraically thus

$$^+100 + {}^-100 = 0.$$

The result, 0, means *neither credit nor debit.*
Similarly for *opposite temperatures.*

Ex. 2. If a body is first heated 10° and then cooled down 8°, its final temperature is 2° above its original temperature; or, stated algebraically,

$$^+10 + {}^-8 = {}^+2.$$

The result, $^+2$, means a *rise* of 2° in temperature.

EXERCISES VIII.

Express algebraically each one of the following statements:

1. $80 gain and $80 loss is equivalent to neither gain nor loss.

2. $75 gain and $60 loss is equivalent to $15 gain.

3. $45 gain and $65 loss is equivalent to $20 loss.

4. If a man travels 100 miles due west and then 100 miles due east, he is at his starting place.

5. If a man ascends 1000 feet in a balloon and then descends 200 feet, he is 800 feet above the earth.

6. If a man walks 80 feet to the right and then 100 feet to the left, he is 20 feet to the left of his starting point.

7. A rise of 18° in temperature, followed by a fall of 26°, is equivalent to a fall of 8°

CHAPTER II.

THE FOUR FUNDAMENTAL OPERATIONS WITH ALGEBRAIC NUMBER.

ADDITION OF ALGEBRAIC NUMBERS.

1. *The* **Addition** *of two numbers is the process of uniting them into one aggregate.*

The numbers to be added are called **Summands.**

Addition of Numbers with Like Signs.

2. Ex. 1. Add $^+3$ to $^+4$.

The three positive units, $^+3$, when added to the four positive units, $^+4$, give an aggregate of *four plus three,* or *seven,* positive units. That is,
$$^+4 + {^+3} = {^+(4+3)} = {^+7}.$$
In like manner,

Ex. 2. $^-4 + {^-3} = {^-(4+3)} = {^-7}.$

These examples illustrate the following method of adding two numbers with like signs:

Add arithmetically their absolute values, and prefix to the sum their common sign of quality.

Addition of Numbers with Unlike Signs.

3. Ex. 1. Add $^-2$ to $^+5$.

The two negative units, $^-2$, when added to the five positive units, $^+5$, *cancel two of the five positive units.* There remain then *five minus two,* or *three,* positive units. That is,
$$^+5 + {^-2} = {^+(5-2)} = {^+3}.$$

Ex. 2. Add $^+2$ to $^-5$.

The two positive units, $^+2$, when added to the five negative units, $^-5$, cancel two of the five negative units. There remain then *five minus two*, or *three*, negative units.

That is, $\qquad ^-5 +{}^+2 = {}^-(5-2) = {}^-3.$

Observe that in both examples *the sum is of the same quality as the number which has the greater absolute value. Also, that the absolute value of the sum is obtained by subtracting the less absolute value, 2, from the greater, 5.*

These examples illustrate the following method of adding two numbers with unlike signs:

Subtract arithmetically the less absolute value from the greater. To that remainder prefix the sign of quality of the number which has the greater absolute value.

The examples given in Ch. I., Art. 34, are concrete illustrations of the preceding principles.

4. Observe that a *positive* number *increases* a number to which it is added, while a *negative* number *decreases*.

EXERCISES I.

Add:

1.	2.	3.	4.	5.	6.
$^+3$	$^-4$	$^+7$	5	$^+11$	$^-19$
$^+5$	$^-8$	$^+2$	$^-9$	$^+20$	$^-4$

7.	8.	9.	10.	11.	12.
$^+7$	$^-7$	$^-8$	$^+11$	$^-19$	$^+45$
$^-5$	$^+5$	$^+8$	$^-16$	$^+32$	$^-22$

SUBTRACTION OF ALGEBRAIC NUMBERS.

5. Subtraction *is the inverse of addition.* In addition two numbers are given, and it is required to find their sum, as in $^+9 + {}^+2 = {}^+11$.

In subtraction the sum of two numbers and one of them are given, and it is required to find the other number, as in

$$^+11 - {}^+2 = ({}^+9 + {}^+2) - {}^+2 = {}^+9.$$

SUBTRACTION OF ALGEBRAIC NUMBERS.

That is, *if from the sum of two numbers either of the numbers be subtracted, the remainder is the other number.*

In general, $(a + b) - a = b.$

6. Ex. 1. A man's net profits last year were 1200 dollars. This year his income is 150 dollars less, and his expenditures are the same. What are his net profits this year?

*To take away 150 dollars income is **equivalent to** adding 150 dollars expenditures.*

If net profits and income be taken positively, and expenditures negatively, the last statement, expressed algebraically, is

$$^{+}1200 - {}^{+}150 = {}^{+}1200 + {}^{-}150.$$

Ex. 2. A man's net profits last year were 1200 dollars. This year his income is the same and his expenditures are 150 dollars less. What are his net profits for this year?

To take away 150 dollars expenditures is equivalent to adding 150 dollars profits.

The algebraic statement of this relation is

$$^{+}1200 - {}^{-}150 = {}^{+}1200 + {}^{+}150.$$

These examples illustrate the following principle:

To subtract one number from another number, reverse the sign of quality of the subtrahend, and add.

E.g., $\quad {}^{+}2 - {}^{+}3 = {}^{+}2 + {}^{-}3, = {}^{-}1. \quad {}^{-}2 - {}^{+}3 = {}^{-}2 + {}^{-}3 = {}^{-}5.$

$\quad\quad\ {}^{+}2 - {}^{-}3 = {}^{+}2 + {}^{+}3, = {}^{+}5. \quad {}^{-}2 - {}^{-}3 = {}^{-}2 + {}^{+}3 = {}^{+}1.$

7. It is important to notice that the preceding examples do not prove this principle. The following examples illustrate a method of proof which may be used ·

Ex. 1. Subtract ${}^{+}5$ from ${}^{+}7$.

In ${}^{+}7 - {}^{+}5$, the minuend, ${}^{+}7$, is to be expressed as the sum of two numbers, *one of which is* ${}^{+}5$. Since ${}^{-}5 + {}^{+}5 = 0$, we may write

$$^{+}7 = {}^{+}7 + {}^{-}5 + {}^{+}5.$$

That is, $^+7$ may be regarded as the sum of two numbers, one of which is $^+7 + {}^-5$, and the other is $^+5$. Therefore, by definition of subtraction,

$$^+7 - {}^+5 = [(^+7 + {}^-5) + {}^+5] - {}^+5$$
$$= {}^+7 + {}^-5 = {}^+2,$$

That is, *to subtract $^+5$ is equivalent to adding $^-$5.*

Ex. 2. Subtract $^-5$ from $^+\iota$.

We have $\quad ^+7 - {}^-5 = [(^+7 + {}^+5) + {}^-5] - {}^-5$
$$= {}^+7 + {}^+5 = {}^+12,$$

That is, *to subtract $^-5$ is equivalent to adding $^+5$.*

8. We thus see that every operation of subtraction is equivalent to an operation of addition. On this account it is convenient to speak of a chain of additions and subtractions as an **Algebraic Sum**.

EXERCISES II.

Subtract:

1.	2.	3.	4.	5.	6.
$^+8$	$^+5$	$^+9$	$^+3$	$^-7$	$^-2$
$^+5$	$^+8$	$^+3$	$^+9$	$^-2$	$^-7$

7.	8.	9.	10.	11.	12.
$^-8$	$^-9$	$^+5$	$^+11$	$^-19$	$^+99$
$^-8$	$^-5$	$^-9$	$^-16$	$^+7$	$^-43$

MULTIPLICATION OF ALGEBRAIC NUMBERS.

9. In multiplication, the multiplicand and multiplier are called **Factors** of the product.

10. In ordinary Arithmetic, multiplication by an integer is defined as an abbreviated addition. Thus,

$$4 \times 3 = 4 + 4 + 4;$$

that is, the number 4 is taken three times **as a summand**.

But $$3 = 1 + 1 + 1.$$

We thus see that the product 4×3 is obtained from 4 just as 3 is obtained from the unit, 1.

We are thus naturally led to the following definition of multiplication:

The product is obtained from the multiplicand just as the multiplier is obtained from the positive unit.

11. There are two cases to be considered in the multiplication of algebraic numbers.

(i.) **The Multiplier Positive.** — Ex. 1. Multiply $^+4$ by $^+3$.

By the definition of multiplication, the product,

$$^+4 \times {}^+3,$$

is obtained from $^+4$ just as $^+3$ is obtained from the positive unit. But
$$^+3 = {}^+1 + {}^+1 + {}^+1.$$

Consequently the required product is obtained by taking $^+4$ *three times as a summand,* or

$$^+4 \times {}^+3 = {}^+4 + {}^+4 + {}^+4 = {}^+(4 + 4 + 4) = {}^+(4 \times 3) = {}^+12.$$

Ex. 2. Multiply $^-4$ by $^+3$.

By the definition of multiplication, we have

$$^-4 \times {}^+3 = {}^-4 + {}^-4 + {}^-4 = {}^-(4 + 4 + 4) = {}^-(4 \times 3) = {}^-12.$$

(ii.) **The Multiplier Negative.** — Ex. 3. Multiply $^+4$ by $^-3$.

By the definition of multiplication, the product,

$$^+4 \times {}^-3,$$

is obtained from $^+4$ just as $^-3$ is obtained from the positive unit. But
$$^-3 = {}^-1 + {}^-1 + {}^-1 = - {}^+1 - {}^+1 - {}^+1;$$

that is, ⁻3 is obtained *by subtracting* the positive unit, ⁺1, *three times in succession from* 0. Consequently, the required product is obtained *by subtracting the multiplicand*, ⁺4, *three times in succession from* 0; or,

$$^+4 \times {}^-3 = -{}^+4 - {}^+4 - {}^+4 = +{}^-4 + {}^-4 + {}^-4 = {}^-(4 \times 3).$$

Ex. 3. Multiply ⁻4 by ⁻3.

By the definition of multiplication, we have

$$^-4 \times {}^-3 = -{}^-4 - {}^-4 - {}^-4 = +{}^+4 + {}^+4 + {}^+4 = {}^+(4 \times 3).$$

12. These examples illustrate the following **Rule of Signs** for **Multiplication**:

The product of two numbers having like signs is positive; and the product of two numbers having unlike signs is negative. Or, stated symbolically,

$$^+a \times {}^+b = {}^+(ab), \qquad {}^-a \times {}^+b = {}^-(ab),$$
$$^-a \times {}^-b = {}^+(ab). \qquad {}^+a \times {}^-b = {}^-(ab).$$

EXERCISES III.

Multiply·

1.	2.	3.	4.	5.	6.
+2	+2	-2	-2	+7	-8
+5	-5	+5	-5	+9	-5

7.	8.	9.	10.	11.	12.
-9	+6	-11	+12	-15	+15
+4	-3	- 3	+ 4	+ 5	- 5

DIVISION OF ALGEBRAIC NUMBERS.

13. Division *is the inverse of multiplication*. In multiplication two factors are given, and it is required to find their product. In division the product of two factors and one of them are given, and it is required to find the other factor.

E.g., since $\qquad ^-28 = {}^-4 \times {}^+7,$

therefore, $\qquad ^-28 \div {}^+7 = {}^-4,$ and $^-28 \div {}^-4 = {}^+7.$

THE SIGNS OF QUALITY AND OPERATION.

14. From the definition of division we infer the following principle:

If the product of two numbers be divided by either of them, the quotient is the other number.

15. Since $^+a \times {}^+b = {}^+(ab)$, therefore $^+(ab) \div {}^+a = {}^+b$;

since $^-a \times {}^+b = {}^-(ab)$, therefore $^-(ab) \div {}^-a = {}^+b$;

since $^-a \times {}^-b = {}^+(ab)$, therefore $^+(ab) \div {}^-a = {}^-b$;

since $^+a \times {}^-b = {}^-(ab)$, therefore $^-(ab) \div {}^+a = {}^-b$.

From these equations, we derive the following **Rule of Signs for Division**:

Like signs of dividend and divisor give a positive quotient; unlike signs of dividend and divisor give a negative quotient.

E.g., $\quad\quad {}^+8 \div {}^+2 = {}^+4; \ {}^-8 \div {}^-2 = {}^+4;$

$\quad\quad\quad\quad {}^+8 \div {}^-2 = {}^-4; \ {}^-8 \div {}^+2 = {}^-4.$

EXERCISES IV.

Divide:

1.	2.	3.	4.	5.
$^+3\overline{){}^+9}$	$^+3\overline{){}^-9}$	$^-3\overline{){}^+9}$	$3\overline{){}^-9}$	$^-4\overline{){}^+12}$
6.	**7.**	**8.**	**9.**	**10.**
$6\overline{){}^-18}$	$^+5\overline{){}^-15}$	$7\overline{){}^-21}$	$^+9\overline{){}^+18}$	$^+7\overline{){}^-28}$

ONE SET OF SIGNS FOR QUALITY AND OPERATION.

16. Most text-books of Algebra use the one set of signs, $+$ and $-$, to denote both *quality* and *operation*. We shall in subsequent work follow this custom. For the sake of brevity, the sign $+$ is usually omitted when it denotes *quality;* the sign $-$ is never omitted.

Thus, instead of $^+2$, we shall write $+2$, or 2;

instead of $^-2$, we shall write -2.

17. We have used the double set of signs hitherto in order to emphasize the difference between *quality* and *operation*. It should be kept clearly in mind that the same distinction still exists.

We now have

$^+3 + {^+2} = +3 + (+2) = 3 + 2$, omitting the signs of *quality*, $+$;

$^+3 + {^-2} = +3 + (-2)$, wherein $+$ denotes *operation*, and $-$ denotes *quality*.

$^+3 - {^+2} = +3 - (+2) = 3 - 2$, omitting the signs of *quality*, $+$;

$^+3 - {^-2} = +3 - (-2)$, wherein the first sign, $-$, denotes *operation*, the second sign, $-$, denotes *quality*.

18. In the chain of operations

$$(+2) + (-5) - (+2) - (-11)$$

the signs within the parentheses denote *quality*, those without denote *operation*. That expression reduces to

$$(+2) - (+5) - (+2) + (+11),$$

or $$2 - 5 - 2 + 11,$$

dropping the sign of *quality*, $+$.

In the latter expression all the signs denote *operation*, and the numbers are all *positive*.

19. The following examples illustrate the double use of the signs $+$ and $-$

Ex. 1. $^+4 + {^+3} = +4 + (+3) = 4 + 3 = 7$.

Ex. 2. $^-5 + {^+2} = -5 + (+2) = -5 + 2 = -3$.

Ex. 3. $^+7 - {^-5} = +7 - (-5) = +7 + (+5) = 7 + 5 = 12$.

Ex. 4. $^-4 \times {^+3} = -4 \times (+3) = -4 \times 3 = -12$.

Ex. 5. $^-4 \times {^-3} = -4 \times (-3) = 12$.

20. In a succession of additions, subtractions, multiplications, and divisions, the multiplications and divisions are first to be performed, and then the additions and subtractions.

E.g., $2 \times (-3) + (-4) \times 5 = -6 + (-20) = -26$.

THE SIGNS OF QUALITY AND OPERATION.

When a different order of performing the operations is proposed, the required order must be indicated by the insertion of parentheses.

E.g., $2 \times [-3 + (-4)] \times 5 = 2 \times (-7) \times 5 = -70$.

EXERCISES V.

In the expressions in Exx. 1–4, which signs denote quality and which operation:

1. $+7 - (+3) + (-4)$.
2. $-9 - (-8) + (+5)$.
3. $(+2) \times (-3) + (-6)$.
4. $(-9) \times (-4) \div (+12)$.

5–8. Find the values of the expressions in Exx. 1–4.

Find the values of the expressions in Exx. 9–20, first changing them into equivalent expressions in which there is only the one set of signs $+$ and $-$:

9. $^+3 + ^+5$.   10. $^+9 - ^+6$.   11. $^+6 - ^+9$.   12. $^-7 + ^-4$.
13. $^-7 - ^+4$.   14. $^-8 - ^-2$.   15. $^+5 \times ^+2$.   16. $^+5 \times ^-2$.
17. $^-5 \times ^-2$.   18. $^+10 \div ^+2$.   19. $^+10 \div ^-2$.   20. $^-10 \div ^-2$.

Simplify the following expressions:

21. $8 - 5$.
22. $5 - 8$.
23. $-9 - 12$.
24. $11 - 4$.
25. $4 - 11$.
26. $-15 + 15$.
27. 10×4.
28. -10×4.
29. $10 \times (-4)$.
30. $-10 \times (-4)$.
31. $27 \div 3$.
32. $-27 \div (-3)$.
33. $27 \div (-3)$.
34. $27 \div 3$.
35. $-36 \div (-4)$.

When $a = 12$, $b = 4$, $c = -3$, $d = -6$, what is the value of

36. $a + b + c$.
37. $a + b - c$.
38. $a - b + c$.
39. $a - (b + c)$.
40. $a - b - c$.
41. $a - (b - c)$.
42. abc.
43. $(ab) \div c$.
44. $a \div (bc)$.
45. $a + b \times c$.
46. $abcd$.
47. $(ab) \div (cd)$.
48. $(abc) \div d$.
49. $ab + cd$.
50. $a + b - d \div c$.

51. A's assets are $3100, and B's are $2700. How much do A's assets exceed B's, taking assets positively?

32   ALGEBRA.   [Ch. II

52. A's assets are $2000, and B owes $500. How much does A's fortune exceed B's, taking assets positively?

53. The temperature in a room is 65° above zero, and out of doors it is 10° above zero. How much higher is the temperature in the room than out of doors, taking degrees above zero positively?

54. The temperature in a room is 68° above zero, and out of doors it is 3° below zero. How much higher is the temperature in the room than out of doors, taking degrees above zero positively?

PARENTHESES.

21. The **Terms** of an algebraic sum are the *additive* and *subtractive* parts of the sum.

E.g., the terms of $2-5-2+11$ are $+2, -5, -2, +11$.

The **Sign of a Term** is its sign $+$ or $-$

A **Positive Term** is one whose sign is $+$; as $+2$.

A **Negative Term** is one whose sign is $-$; as -5.

Removal of Parentheses.

22. We have $\quad 9+(5+6)=9+5+6,$

since to add the sum $5+6$ is equivalent to adding successively the single numbers of that sum.

Again, $\quad 9+(5-6)=9+[5+(-6)],$

since to *add* -6 is equivalent to *subtracting* 6.

Therefore, removing brackets,

$$9+(5-6)=9+5+(-6), =9+5-6.$$

It is important to notice that if the first term within the parentheses has no sign, the sign $+$ is understood.

The above example illustrates the following principle:

(i.) *When the sign of addition, $+$, precedes parentheses, they may be removed, and the signs, $+$ and $-$, within them be left unchanged;* that is,

$$N+(+a+b)=N+a+b,$$
$$N+(+a-b)=N+a-b, \text{ etc.}$$

23. We also have
$$9-(5+6)=9-5-6,$$
since to subtract the sum $5+6$ is equivalent to subtracting successively the single numbers of that sum.

Again, $\qquad 9-(5-6)=9-[5+(-6)],$
since to *add* -6 is equivalent to *subtracting* 6.

Therefore, removing brackets,
$$9-(5-6)=9-5-(-6), =9-5+6.$$
This example illustrates the following principle:

(ii.) *When the sign of subtraction, $-$, precedes parentheses, they may be removed, if the signs within them be reversed from $+$ to $-$, and from $-$ to $+$*; that is,
$$N-(+a+b)=N-a-b,$$
$$N-(+a-b)=N-a+b, \text{ etc.}$$

Observe that the sign before the parentheses affects each term within them.

Insertion of Parentheses.

24. The insertion of parentheses is the converse of the process of removing them.

(i.) *An expression may be inclosed within parentheses preceded by the sign $+$, if the signs of the terms inclosed remain unchanged.*

E.g., $\qquad 7-5+3-4=7+(-5+3-4),$
$$=7-5+(3-4).$$

(ii.) *An expression may be inclosed within parentheses preceded by the sign $-$, if the signs of the terms inclosed be reversed, from $+$ to $-$ and from $-$ to $+$.*

E.g., $\qquad 7-5+3-4=7-(5-3+4),$
$$=7-5-(-3+4).$$

25. The principle for inserting parentheses enables us to group successive terms in addition and subtraction.

We thus see that the algebraic sum of three or more numbers is the same in whatever way successive numbers are grouped.

EXERCISES VI.

Find the value of each of the following expressions, first removing parentheses:

1. $8 + (4 + 1)$.
2. $8 + (4 - 1)$.
3. $9 + (6 + 2)$.
4. $9 + (-6 + 2)$.
5. $9 - (5 + 3)$.
6. $9 - (5 - 3)$.
7. $9 - (-5 + 3)$.
8. $9 - (7 - 6)$.
9. $9 - (-7 + 6)$.
10. $16 + (8 - 5 + 1)$.
11. $16 - (8 - 5 + 1)$.
12. $25 - (6 - 9 + 5)$.

Insert parentheses in the expression $9 - 7 + 3 - 5$.

13. To inclose the last two terms, preceded by the sign $+$; preceded by the sign

14. To inclose the last three terms, preceded by the sign $+$; preceded by the sign

Order of Successive Operations.

26. It is proved in School Algebra, Ch. II., that the order in which successive additions and subtractions, and successive multiplications and divisions are performed can be changed. Thus,

$$8 - 3 + 2 - 5 = 8 + 2 - 3 - 5 = 10 - 8 = 2, \qquad \text{(i.)}$$
$$25 \times 27 \times 4 = 25 \times 4 \times 27 = 100 \times 27 = 2700, \qquad \text{(ii.)}$$
$$75 \times 29 \div 25 = 75 \div 25 \times 29 = 3 \times 29 = 87. \qquad \text{(iii.)}$$

In thus changing the order of the operations, it is important to *carry the symbol of operation with the number.*

27. By the methods of the preceding article, we secure the following advantages:

In a succession of additions and subtractions, add the positive terms separately, then the negative terms, and unite the results. as in (i.).

In a succession of multiplications and divisions, we may, by changing the order of the operations, frequently simplify the work. as in (ii.) and (iii.).

EXERCISES VII.

Find the value of each of the following expressions:

1. $3 - 4 + 9 - 6 + 2$.
2. $-5 + 7 - 15 + 19 - 6$.
3. $13 - 8 + 6 - 3 - 5$.
4. $2 - 17 - 3 + 18 - 9$.

Find, in the most convenient way, the value of each of the following expressions:

5. $96 - 110 + 4$.
6. $33\frac{1}{3} - 25 + 66\frac{2}{3}$.
7. $995 + 997 + 5 + 3$.
8. $99 + 97 + 93 + 1 + 7 + 3$.
9. $-25 \times 17 \times 4$.
10. $2\frac{1}{2} \times (-38) \times 4$.
11. $-39 \times 3 \times 33\frac{1}{3}$.
12. $-25 \times 12 \div 5$.
13. $33\frac{1}{3} \div (-20) \times 3$.
14. $10 \div (-34) \times 17$.

POSITIVE INTEGRAL POWERS.

28. A continued product of equal factors is called a **Power** of that factor.

Thus, 2×2 is called the *second power* of 2, or *2 raised to the second power*; aaa is called the *third power* of a, or *a raised to the third power*.

In general $aaa \cdots$ to n factors is called the nth power of a, or *a raised to the nth power*.

The second power of a is often called the *square* of a, or *a squared*; and the third power of a the *cube* of a, or *a cubed*.

29. The notation for powers is abbreviated as follows:
a^2 is written instead of aa; a^3 instead of aaa;
a^n instead of $aaa \cdots$ to n factors.

30. The **Base** of a power is the number which is repeated as a factor.

E.g., a is the base of a^2, a^3, $\cdots$, a^n.

31. The **Exponent** of a power is the number which indicates how many times the base is used as a factor, and is written to the right and a little above the base.

E.g., the exponent of a^2 is 2, of a^3 is 3, of a^n is n.

The exponent 1 is usually omitted. Thus, $a^1 = a$.

32. The base of a power must be inclosed within parentheses to prevent ambiguity :

(i.) *When the base is a negative number.* Thus,
$(-5)^2 = (-5)(-5) = 25$; while $-5^2 = -(5 \times 5) = -25$.

(ii.) *When the base is a product or a quotient.* Thus,
$$(2 \times 5)^3 = (2 \times 5)(2 \times 5)(2 \times 5) = 1000;$$
while $2 \times 5^3 = 2(5 \times 5 \times 5) = 250$.

Likewise $\left(\dfrac{2}{3}\right)^2 = \dfrac{2}{3} \times \dfrac{2}{3} = \dfrac{4}{9}$, while $\dfrac{2^2}{3} = \dfrac{2 \times 2}{3} = \dfrac{4}{3}$.

(iii.) *When the base is a sum.* Thus,
$$(2+3)^2 = (2+3)(2+3) = 5 \times 5 = 25;$$
while $2 + 3^2 = 2 + 3 \times 3 = 2 + 9 = 11$.

(iv.) *When the base is itself a power.* Thus,
$$(2^3)^2 = 2^3 \times 2^3 = (2 \times 2 \times 2)(2 \times 2 \times 2) = 64,$$
while $2^{3^2} = 2^{3 \times 3} = 2^9 = 2 \times 2 \times 2 \times 2 \times 2 \times 2 \times 2 \times 2 \times 2 = 512$.

EXERCISES VIII.

Express each of the following powers in the abbreviated notation:

1. $a \times a$.   2. 3×3.   3. $2 \times 2 \times 2$.   4. $(-a)(-a)$.
5. $bbbb$.   6. $(-3)(-3)(-3)(-3)$.   7. $-nnnnn$.
8. $2 \times 2 \times 2 \cdots$ to 8 factors.   9. $(-a)(-a)(-a) \cdots$ to 9 factors.
10. $(a+b)(a+b)(a+b)$.
11. $(x+y)(x+y) \cdots$ to 12 factors.

Express each of the following powers as a continued product, and find (in 12–15) its value:

12. 4^5.   13. 5^4.   14. $(-3)^3$.   15. -3^3.
16. xy^3.   17. $(xy)^3$.   18. $-a^4$.   19. $(-a)^4$.

Write

20. Three times a.   **21.** a to the third power.

22. The sum of the squares of a and b.

23. The square of the sum of a and b.

Properties of Positive Integral Powers.

33. (i) *All (even and odd) powers of positive bases are positive.*

E.g., $2^3 = 2 \times 2 \times 2 = 8.$ $3^4 = 3 \times 3 \times 3 \times 3 = 81.$

(ii.) *Even powers of negative bases are positive; odd powers of negative bases are negative.*

E.g., $\quad (-2)^4 = (-2)(-2)(-2)(-2) = 16;$

$\quad\quad\quad (-5)^3 = (-5)(-5)(-5) = -125.$

EXERCISES IX.

Find the value of each of the following powers:

1. 4^3.   **2.** 3^4.   **3.** $(-2)^4$.   **4.** -2^4.   **5.** $(-2)^5$.
6. $(-3)^6$.   **7.** $(-3)^3$.   **8.** -3^3.   **9.** $(-x)^6$.   **10.** $(-x)^9$.

Express as powers of 2:

11. 8.   **12.** 32.   **13.** 256.   **14.** 1024.

Express as powers of -3:

15. 9.   **16.** -27.   **17.** -243.   **18.** 729.

Find the value of each of the following expressions:

19. $2^2 + 3^2$.   **20.** $(2+3)^2$.   **21.** $(3^3 - 2^3)$.   **22.** $(3-2)^3$.
23. $(5 \times 2)^2$.   **24.** 5×4^2.   **25.** $2(-3)^3$.   **26.** $[2(-3)]^3$

When $a = 3$, $b = -3$, $c = 2$, find the value of each of the following expressions:

27. a^c.   **28.** b^c.   **29.** $(ab)^c$.   **30.** bc^a.   **31.** $(abc)^c$.
32. $a^2 - b^2 + c^2$.   **33.** $(a - b + c)^2$.   **34.** $[a^2 - b^2 - c^2]^2$.

CHAPTER III.

THE FUNDAMENTAL OPERATIONS WITH INTEGRAL ALGEBRAIC EXPRESSIONS.

DEFINITIONS.

1. An **Integral Algebraic Expression** is an expression in which the *literal* numbers are connected only by one or more of the symbols of operation, $+$, $-$, $\times$, but not by the symbol $\div$.

E.g., $1 + x + x^2$, $5 a^2 b + \frac{2}{3} c d^2$, etc.

2. The word *integral* refers only to the *literal* parts of the expression.

E.g., $a + b$ is *algebraically* integral; but when $a = \frac{1}{2}$, $b = \frac{3}{4}$, we have
$$a + b = \tfrac{1}{2} + \tfrac{3}{4} = 1\tfrac{1}{4}.$$

3. Coefficients. — In a product, any factor, or product of factors, is called the **Coefficient** of the product of the remaining factors.

E.g., in $3\, abc$, 3 is the coefficient of abc, $3\, b$ of ac, etc.

A **Numerical Coefficient** is a coefficient expressed in figures.

E.g., in $-3\, ab$, -3 is the numerical coefficient of ab.

A **Literal Coefficient** is a coefficient expressed in letters, or in letters and figures.

E.g., in $3\, ab$, a is the literal coefficient of $3\, b$, and $3\, a$ of b.

The coefficients $+1$ and -1 are usually omitted.

4. A coefficient must not be confused with **an exponent.**

E.g., $4\, a = a + a + a + a$; while $a^4 = a \times a \times a \times a$.

ADDITION AND SUBTRACTION.

5. The sign $+$, or the sign $-$, preceding a product, is to be regarded as the sign of its numerical coefficient.

Thus, $+3a$ means the product of *positive* 3 by a; $-5x$ means the product of *negative* 5 by x. In particular, $+a$ means the product of *positive* 1 by a, and $-a$ means the product of *negative* 1 by a, unless the contrary is stated.

6. Like or **Similar Terms** are terms which do not differ, or which differ only in their numerical coefficients.

E.g., in the expression $+3a+6ab-5a+7ab$, $+3a$ and $5a$ are like terms; so are $+6ab$ and $+7ab$.

Unlike or **Dissimilar Terms** are terms which are not like.

E.g., $+3a$ and $+7ab$ in the above expression.

7. A **Monomial** is an expression of one term; as a, $-7bc$.

A **Binomial** is an expression of two terms; as $-2a^2+3bc$.

A **Trinomial** is an expression of three terms.

E.g., $\qquad a+b-c, \ -3a^2+7b^3-5c^4.$

A **Multinomial*** is an expression of two or more terms, including, therefore, binomials and trinomials as particular cases.

E.g., $\qquad a+b^2, \ a^2+b-c^3, \ ab+bc-cd-ef.$

ADDITION AND SUBTRACTION.

Addition of Like Terms.

8. Like Terms can be united by addition into a single *like* term.

Just as $\quad 2=1+1$, so $2xy=xy+xy$;

just as $\quad 3=1+1+1$, so $3xy=xy+xy+xy$.

Therefore, just as $\quad 2+3=5$,

so $\qquad 2xy+3xy=(2+3)xy=5xy.$

That is, *to add like terms, add their numerical coefficients and annex to the sum their common literal part.*

---

* The word **Polynomial** is frequently used instead of **Multinomial**.

Ex. 1. Add $-7ab$ to $4ab$.

We have $4ab + (-7ab) = [4 + (-7)]ab = -3ab$.

Ex. 2. Find the sum of $3a$, $-5a$, $8a$, $-4a$.

Uniting the positive terms by themselves, and the negative terms by themselves, we have

$$3a + 8a + (-5a) + (-4a) = [3 + 8 + (-5) + (-4)]a = 2a.$$

Ex. 3. Add ax to bx.

Since the sum of the coefficients of x is $a + b$, we have
$$ax + bx = (a+b)x.$$

EXERCISES I.

Add:

1.	2.	3.		5.	6.
a	$-2b$	$8x$	$-12m$	$3a$	$5x$
$2a$	$-b$	$4x$	$-8m$	$-5a$	$-7x$

7.	8.	9.	10.	11.	12.
$9x^2$	$2ab$	$5xy$	$3a^2$	cx	ax^3
$-4x^2$	$5ab$	$7xy$	$-6a^2$	dx	$-bx^3$

Find the sum of:

13. $x, 3x, 4x, 7x$.   14. $-2a, -5a, -a, -6a$.

15. $5x^2, -4x^2, 6x^2, -3x^2$.   16. $-9mn, 7mn, -4mn, 3mn$.

17. $3ab, 4ab, -7ab, 2ab, -8ab$.

18. $4a^2, -9a^2, 5a^2, -3a^2, -2a^2$.

19. $a+b, -3(a+b), 7(a+b), -8(a+b), 2(a+b)$.

Simplify the following expressions:

20. $a + 2a + 3a$.   21. $a - 2a + 3a$.

22. $ax - bx + cx$.   23. $7x + 3x - x - 5x + 4x$.

24. $15m - 12m + 2m - m$.   25. $a^2 - 2a^2 + 3a^2 - 5a^2$.

ADDITION AND SUBTRACTION.

26. $-m^2 + 3m^2 + 8m^2 - 10m^2$.
27. $ab + 3ab - 2ab - 5ab$.   28. $5xy - 2xy + 6xy - 8xy$.
29. $ax - 3ax - 5ax + 2ax$.   30. $7am + 3am - 2am - 8am$.
31. $3a^2b + 5a^2b - 13a^2b - 2a^2b + 6a^2b$.
32. $25mn - 17mn + 8mn + 9mn - 24mn$.
33. $a^3b + 6a^3b + 11a^3b - 12a^3b + 5a^3b - 12a^3b$.
34. $(a+2b) - 3(a+2b) - 5(a+2b) + 8(a+2b)$.
35. $7(x^2+y^2) + 5(x^2+y^2) - 10(x^2+y^2) + 3(x^2+y^2)$.

Subtraction of Like Terms.

9. Like Terms can be united by subtraction into a single like term.

Just as $\qquad 5 - 2 = 3,$

so $\qquad 5a - 2a = (5-2)a = 3a.$

That is, *to subtract like terms, subtract their numerical coefficients and annex to the remainder their common literal part.*

Ex. 1. Subtract $-5x^2y$ from $-7x^2y$.
We have
$$7x^2y - (-5x^2y) = -7x^2y + 5x^2y = (-7+5)x^2y = -2x^2y.$$

EXERCISES II.

Subtract:

1.	2.	3.	4.	5.	6.
$2a$	$8x$	$-6b$	$-4m$	$3a$	$10b$
a	$3x$	$5b$	$-6m$	$-2a$	$-5b$

7.	8.	9.	10.	11.	12.
$-x$	$2y$	$-6m$	$-15x$	$5x^2$	$4a^3$
$-2x$	$-3y$	$-4m$	$-10x$	x^2	$-6a^3$

13. $10a^2b$ from $12a^2b$.   14. $15xy$ from $10xy$.

15. $-7x^3$ from $8x^3$. **16.** $-3mn$ from $-8mn$.
17. $3(a+b)$ from $-5(a+b)$. **18.** $-7(x-y)$ from $-9(x-y)$.

Addition of Multinomials.

10. Unlike Terms are added by writing them in succession, each preceded by the sign $+$.

Ex. 1. Add $3b$ to $2a$. We have $2a + 3b$.

Ex. 2. Add $-3x^2$ to $2y^2$. We have
$$2y^2 + (-3x^2) = 2y^2 - 3x^2.$$

Such steps as changing $+(-3x^2)$ into $-3x^2$, should be performed mentally.

11. A multinomial consisting of two or more sets of like terms can be simplified by uniting like terms.

Ex. 1. $2a - 3b - 5a + 4b = 2a - 5a - 3b + 4b$
$$= -3a + b.$$

12. If two or more multinomials have common like terms, these terms can be united.

Ex. 1. Add $-2a + 3b$ to $3a - 5b$.
We have $(3a - 5b) + (-2a + 3b) = 3a - 5b - 2a + 3b$,
$$= a - 2b.$$

In adding multinomials, it is often convenient to write one underneath the other, placing like terms in the same column.

Ex. 2. Find the sum of $-4x^2 + 3y^2 - 8z^2$, $2x^2 - 3z^2$, and $y^2 + 5z^2$

We have
$$\begin{array}{r} -4x^2 + 3y^2 - 8z^2 \\ 2x^2 \qquad\qquad 3z^2 \\ 2y^2 + 5z^2 \\ \hline -2x^2 + 5y^2 - 6z^2 \end{array}$$

It is evidently immaterial whether the addition is performed from left to right, or from right to left, since there is no carrying as in arithmetical addition.

ADDITION AND SUBTRACTION.

EXERCISES III.

Add:

1.	2.	3.	4.	5.	6.
x	2	3	5	x	$-x$
1	y	$-a$	x	y	$-y$

7. $5a$ to $-6m$. 8. $-7x$ to $10y$. 9. $-2x$ to $-3y$.
10. x^2 to x. 11. $3a^2$ to a. 12. $-5m^2$ to $-3n^2$.

Simplify the following expressions by uniting like terms:

13. $a + 1 + a - 1$.
14. $3x + 2 - 2x - 1$.
15. $5m + 7 - 2 + m$.
16. $5a + 2b + 3b - 2a - 3a$.
17. $x^2 - 3x + 5x + x^2$.
18. $m^3 + m^2 - 3m^3 - 2m^2 + 2m^3 + m^3$.
19. $a^3 + a^2 + a + 3a^2 - 5a^3 - 6a - 2a + 4a^3 - 4a^2$.
20. $7x^2y - 3xy - 5xy + 8x^2y - 2xy - 10x^2y - 5x^2y$.
21. $2ab - 9xy + 3xy - 5ab - 2ab + 16xy + ab$.
22. $5(m+n) - 2m - 3m - 2(m+n) + 2m$.
23. $-8(x^2 + y^2) - 3x^2 - 3y^2 + 7(x^2 + y^2) + 2x^2 + 2y^2$.

Find the sums of the following expressions:

24. $7a + 3b$, $5a - 4b$, $2b - 3a$, $6b - 4a$.
25. $5m - 6n$, $n - 3m$, $4m - 7n$, $9n - 2m$.
26. $2x - 3y + 4z$, $5x - 6z + 7y$, $z - 3x - 8y$.
27. $3x^2 + x + 1$, $5x^2 - 3x + 6$, $-7x^2 + x - 6$.
28. $5a^2 - 3a + 8$, $-4a^2 + 5a - 6$, $-a - 1$.
29. $2x^2 - 3xy + y^2$, $5xy - 4x^2 - 5y^2$, $3x^2 + 2xy + 7y^2$
30. $3m - 7m^2 + 4$, $3 - m$, $10m^2 - 6$.
31. $6x^2y^2 - 3xy + 8$, $xy + 5$, $-3x^2y^2 - 12$.
32. $2x^3 - 3x^2 + 6x - 1$, $x^2 - 4x + 3$, $2x - 7$.
33. $a^3 - 5a^2b + 6ab^2$, $a^2b - 2ab^2 - 3b^3$, $-4ab^2 + 5b^3$.
34. $7(x^2 - y^2) - 3(x + y) + 9$, $5(x + y) - 6$, $-4(x^2 - y^2) - 2$

Subtraction of Multinomials.

13. Unlike Terms are subtracted by writing them in succession, each preceded by the sign —.

Ex. Subtract $-11\,m$ from $2\,n$. We have
$$2\,n - (-11\,m) = 2\,n + 11\,m.$$

14. If two multinomials have common like **terms**, these terms can be united.

Ex. 1. Subtract $-2\,a + 3\,b$ from $3\,a - 5\,b$.

We have $(3\,a - 5\,b) - (-2\,a + 3\,b) = 3\,a - 5\,b + 2\,a - 3\,b,$
$$= 5\,a - 8\,b.$$

Ex. 2. Subtract $2\,x^2 - 6\,x - 3$ from $3\,x^2 - 5\,x + 1$.

Changing mentally the signs of the **terms of the subtrahend**, and adding, we have
$$\begin{array}{r} 3\,x^2 - 5\,x + 1 \\ 2\,x^2 - 6\,x - 3 \\ \hline x^2 + x + 4 \end{array}$$

Ex. 3. Subtract $2\,x^2 \quad 3\,z^2$ from $-4\,x^2 + 3\,y^2$, and from the result subtract $2\,y^2 + 5\,z^2$.

When several multinomials are to be subtracted in succession, the work is simplified by writing them with the signs of the terms already changed. We then have
$$\begin{array}{r} -4\,x^2 + 3\,y^2 \\ 2\,x^2 + 3\,z^2 \\ -2\,y^2 - 5\,z^2 \\ \hline -6\,x^2 + y^2 - 2\,z^2 \end{array}$$

EXERCISES IV.

Subtract:

1.	2.	3.	4.	5.	6.
a	$2\,x$	$4\,x^2$	$-5\,mn$	$7\,xy$	$3\,a^2b$
b	$-3\,y$	$2\,x$	$-2\,n^2$	$-2\,xz$	$-5\,ab^2$

ADDITION AND SUBTRACTION.

7. $2a + 3b$ from $3a + 4b$.   8. $4a + 6b$ from $5a + 9b$.
9. $5x - 6y$ from $6x - 7y$.   10. $3x - 9y$ from $4x - 8y$.
11. $4m + n$ from $5m - n$.   12. $2a - 3b$ from $3a + 5b$.
13. $3x^2 + 2x$ from $-2x^2 - 3x$.   14. $m^3 + n$ from $m^3 + 6m$.
15. $2a - 3b + 5c$ from $4a - 6b + 8c$.
16. $3x^2 - 5y^2 + 7z^3$ from $5x^2 - 2y^2 + 3z^2$.
17. $5ab - 3ac - 6bc$ from $8ab + 5ac - 7bc$.
18. $4a^2 + 5b^2 - 7c^2$ from $3a^2 + 4b^2 - 8c^2$.
19. $3m^2n^2 - 5$ from $3mn - 2$.
20. $5xy + 1$ from $x^2y^2 + 4xy + 2$.
21. $x^2y^2 - 3xy - 5$ from $-x^2y^2 - 2$.
22. $x^2 + 3x + 1$ from $4x^2 + x + 1$.
23. $-3m^2 - 2m + 5$ from $2m^2 + 3m$.
24. $2a^2 - 5a + 3$ from $4a^2 - 9a - 2$.
25. $4a^2 - 5ab + 6b^2$ from $5a^2 + ab - 3b^2$.
26. $x^3 - 3x^2 + 2x + 1$ from $2x^3 + 5x + 2$.
27. $2x^2 - 3x - 5$ from $x^3 + 3x^2 - 4x - 6$.
28. $5a^4 - 3a^3 + a - 1$ from $5a^2 - 4a + 6$.
29. $2(a + b) + c$ from $5(a + b) - 2c$.
30. $3(a - b) - 5a - 3b$ from $5(a - b) - 7a - 5b$.

31. From the sum of $2x - 3y + 4z$ and $3x + 5y - 7z$ subtract $4x + 3y - 4z$.

32. From $4a^2 - 3ab - 5b^2$ minus $5a^2 + 2ab - 9b^2$ subtract $3b^2 - 7ab - 2a^2$.

33. From $3x^2 + 7x - 5$ minus $5x^2 + 9x - 1$ subtract $4x^2 - 2x - 6$.

Subtract:

34. $ax + by$ from $bx + ay$.   35. $bx - by$ from $cx - dy$.
36. $ax - by + cz$ from $bx + cy - az$.

PARENTHESES.

15. The use of parentheses has been briefly discussed in Ch. II., Arts. 21–25. It is frequently necessary to employ more than two sets of parentheses, and to distinguish them the following forms are used:

Parentheses, ( ); Brackets, [ ]; Braces, { }.

A **Vinculum** is a line drawn over an expression, and is equivalent to parentheses inclosing it.

E.g., $(a+b)(c-d) = \overline{a+b} \cdot \overline{c-d}$.

16. The principles given in Ch. II., Arts. 21–23, are to be applied successively when several sets of parentheses are to be removed from a given expression.

17. In removing parentheses we may begin either with the inmost or with the outmost.

The following examples will illustrate the method of removing parentheses, beginning with the inmost:

Ex.
$$4a - \{3a + [2a - (a-1)]\}$$
$$= 4a - \{3a + [2a - a + 1]\}$$
$$= 4a - \{3a + a + 1\}$$
$$= 4a - 4a - 1 = -1.$$

When, in such examples, we come to one of a pair of parentheses, (, or [, or {, we must look for the other of like form. We then treat all that is contained in each pair as a whole.

EXERCISES V.

Simplify each of the following expressions:

1. $587 + (13 + 768)$.
2. $(343 + 14) + 986$.
3. $682 + (400 - 82)$.
4. $298 - (48 - 350)$.
5. $7a - 8a - (5a - a + 3a)$.
6. $7a - (8a - 5a) - (a + 3a)$.
7. $7a - [8a - (5a - a) + 3a]$.
8. $7a - [8a - (5a - a + 3a)]$.
9. $7a - \{8a - [5a - (a + 3a)]\}$.

10. $7x - (8y + 7x)$.
11. $5x - (3x - 2)$.
12. $x - (x - y + z)$.
13. $x - (x - y - z)$.
14. $x - 3y - (3x - 6y) + (4x - 3y)$.
15. $3a + 5b - (6a - 7b) - (11b - 3a)$.
16. $x - [2y + (3x - 4y) - 2x]$.
17. $a - [b - 2a - (3b - 4a) - (2a - b)]$.
18. $5a - [3b - (4c - 4a) - (2b - 3c)]$.
19. $5a^2 - [4a^2 - (4 - 3a)] - (3 - 2a)$.
20. $7a - \{4b - [(2a - 7b) - (6a - 9b)]\}$.
21. $9 - [4m^2 - 5m - (3m^2 - \overline{3m - 8})]$.
22. $3x - 3z - [-(4y - 2x) - (3z - \overline{2y + x})]$.
23. $2a - \{3b - [6c - (5a + \overline{c - 6b}) + (4a - \overline{3b + 5c})]\}$.

EQUATIONS AND PROBLEMS.

18. Ex. Find the value of x from $2x - 5 = 7 + x$.

Adding 5 to both members of the equation, we obtain:
$$2x - 5 + 5 = 7 + 5 + x;$$
or, since $-5 + 5 = 0$, $\quad 2x = 7 + 5 + x$.

Subtracting x from both members of the last equation, we have:
$$2x - x = 7 + 5 + x - x;$$
or, since $x - x = 0$, $\quad 2x - x = 7 + 5$. $\hfill (1)$

Uniting terms, $\quad x = 12$.

Check: $2 \times 12 - 5 = 7 + 12$, or $24 - 5 = 7 + 12$, or $19 = 19$.

19. Observe that equation (1) could have been obtained directly from the given equation by transferring the term -5, with sign changed, to the second member, and the term $+x$, with sign changed, to the first member.

That is, *any term may be transferred from one member of an equation to the other, if its sign be reversed from + to −, or from − to +.*

20. Ex. Find the value of x from the equation $x-3=8-3$.

Adding 3 to both members of the equation, we obtain:
$$x-3+3=8-3+3;$$
or, since $-3+3=0, \quad x=8.$

Check: $\quad 8-3=8-3,$ or $5=5.$

Observe that this step is equivalent to dropping the common term -3 from both members.

That is, *the same term, or equal terms, may be dropped from both members of an equation.*

This step is called *cancellation of equal terms*.

21. These examples illustrate the following method:

Transfer all the terms containing the unknown number to one member of the equation, usually to the first member, and all the terms containing known numbers to the other member.

Unite like terms.

Divide both members by the coefficient of the unknown number.

Check *by substituting the value thus obtained in the given equation.*

22. Pr. A boy being asked his age, replied: "If 10 is added to twice the number of years in my age the sum will be 40." How old was the boy?

Let x stand for the number of years in his age.

Then $2x$ stands for twice that number of years.

The problem states,

in *verbal* language: *Twice the number of years in the boy's age plus* 10 *is equal to* 40;

in *algebraic* language: $\quad 2x+10=40.$

Transferring 10, $\quad\quad 2x=30.$

Dividing by 2, $\quad\quad x=15.$

The boy was 15 years old.

Check: $\quad 2 \times 15 + 10 = 40,$ or $30+10=40,$ or $40=40.$

EQUATIONS AND PROBLEMS.

EXERCISES VI.

Find the value of x in each of the following equations:

1. $x + 1 = 2$.
2. $x - 2 = 3$.
3. $x - 5 = 4$.
4. $x + 8 = 5$.
5. $x - 3 = -6$.
6. $3x = 2x + 5$.
7. $4x - 6 = 3x$.
8. $2x + 1 = 11$.
9. $5x - 6 = 9$.
10. $6x + 7 = 25$.
11. $7x = 3x + 8$.
12. $9x - 8 = 7x$.
13. $3x + 2x - 21 = 4$.
14. $8x = 11 - 5x + 2x$.
15. $7x - 9 = 3x + 31$.
16. $8x - 5 = 17 - 3x$.
17. $10x + 27 = 6 + 17x$.
18. $9x - 7 = 13x - 35$.
19. $17 - 5x = 3x - 23$.
20. $14 - 8x = 19 - 3x$.
21. $98 - 9x = 4x + 7$.
22. $15x - 73 = 51x + 35$.
23. $7x - (3x - 2) = 10$.
24. $3 - (5x - 6) = -6$.
25. $5x - (x + 9) = 2 - (3 - 2x)$.

Solve each of the following problems:

26. If 12 is added to a number, the sum will be 23. What is the number?

27. If 19 is subtracted from a number, the remainder will be 31. What is the number?

28. A merchant's capital is increased by $1200. It then amounts to $3500. What was his capital at first?

29. A man, after spending $18, has $70 left. How much money had he at first?

30. A father is 23 years older than his son. The sum of their ages is 45 years. How old is each?

31. A has $50 less than B, and altogether they have $230. How many dollars has each?

32. At an election A received 125 votes more than B. If the number of votes cast for both was 769, how many votes did each receive?

33. A pole 29 feet long is divided into two parts, so that one part is 7 feet shorter than the other. What is the length of each part?

34. In a number of 2 digits, the tens' digit exceeds the units' digit by 4. If the sum of the digits is 10, what is the units' digit? What is the number?

35. What is the number next greater than 5? Next less? Next greater than x? Next less? Next greater than $x+1$?

36. The sum of two consecutive numbers is 29. What are the numbers?

37. The sum of three consecutive numbers is 42. What are the numbers?

38. The difference between two numbers is 9, and the smaller number is 7. What is the greater number? If the greater is x, what is the smaller number?

39. The difference between two numbers is 13, and their sum is 77. What are the numbers?

40. The difference between two numbers is 22, and their sum is equal to four times their difference. What are the numbers?

41. A father is 30 years older than his son. If four times the son's age is equal to the sum of their ages, how old is each?

42. The length of a room is twice the breadth. If the length is 12 feet more than the breadth, what are the dimensions of the room?

43. A man, being asked the time, replied: "If 14 is subtracted from three times the hour it now is, the remainder will be the hour." What was the hour?

44. Twice a number exceeds 27 by as much as 27 exceeds the number. What is the number?

45. If three times a number exceeds 17 by as much as 19 exceeds the number, what is the number?

46. A has $\$124$ and B has $\$98$. How many dollars must A give B in order that they may have equal amounts?

47. A has 1 more than twice as many pencils as B. If he were to give B 3 pencils, they would have the same number. How many has each?

48. In a company, the number of women was three times the number of men, and the number of children was equal to the number of men and women. If the number of women and children exceeded the number of men by 120, how many of each were in the company?

49. A pole is divided into three parts. The second is twice as long as the first, and the third is 8 feet longer than the first. The length of the pole is equal to the excess of 53 feet over the smallest part. What are the lengths of the parts, and the length of the pole?

50. In a number of two digits, the units' digit is twice the tens' digit. The number is equal to 16 more than twice the units' digit. What is the number?

MULTIPLICATION.

Products of Powers.

23. Ex. 1. $a^3 \times a^4 = (aaa)(aaaa) = aaaaaaa = a^7 = a^{3+4}$

Ex. 2. $xx^2x^3 = x(xx)(xxx) = xxxxxx = x^6 = x^{1+2+3}$

These examples illustrate the following principle:

The exponent of the product of two or more powers of the same base is the sum of the given exponents; or stated symbolically,

$$a^m a^n = a^{m+n}; \quad a^m a^n a^p = a^{m+n+p}; \text{ etc.}$$

EXERCISES VII.

Express each of the following products as a single power:

1. $2^2 \times 2$. 2. $3^3 \times 3$. 3. 5×5^3. 4. $-3(-3)^2$.
5. $(-6)^3(-6)^2$. 6. $2^3(-2)^4$. 7. $5^4(-5)^6$. 8. $-2^3(-2)^3$

9. $a^2 \cdot a^3$. 10. $m^3 \cdot m^6$. 11. $x^3 \cdot x^7$.
12. $(-x)^2(-x)^3$. 13. $a^2 \cdot a^3 \cdot a^4$. 14. $x \cdot x^6 \cdot x^5$.
15. $-a^3(-a)^2(-a)^5$. 16. $(a+b)^2(a+b)^3$. 17. $(a-b)^5(a-b)^4$.

Degree.

24. An integral term which is the product of n letters is said to be of the *nth degree*.

Thus, the **Degree of an Integral Term** is indicated by the sum of the exponents of its literal factors.

E.g., $3\,ab$ is of the *second* degree; $2\,x^2y, = 2\,xxy$, is of the third degree.

The **Degree of a Multinomial** is the degree of that term which is of highest degree.

E.g., the degree of $x^2y + xy^3 - x^2y^3z$ is the degree of x^2y^3z; *i.e.*, the *sixth*.

25. If the terms of a multinomial be arranged so that the exponents of some one letter increase, or decrease, from term to term, the multinomial is said to be arranged to *ascending*, or *descending*, powers of that letter. The letter is called the *letter of arrangement*.

E.g., $x^4 - 3\,x^3y^2 + 2\,x^2y + xy^3$ is arranged to *descending* powers of x, which is then the letter of arrangement; or, when written $x^4 + 2\,x^2y - 3\,x^3y^2 + xy^3$, to *ascending* powers of y, which is then the letter of arrangement.

EXERCISES VIII.

What is the degree of $2\,a^3b^2x^4$,

1. In a? **2.** In x? **3.** In a, b, and x?

4. Arrange $2\,x - 5\,x^3 + 7 - 2\,x^4 + 2\,x^2 + x^5$ to descending powers of x; to ascending powers of x.

5. Arrange $3\,y - 7\,xy^3 + 5\,x^3y^2 + 4\,x^2y^4$ to ascending powers of x; to ascending powers of y.

Multiplication of Monomials by Monomials.

26. Ex. 1. $\quad 3a \times 5b = 3 \times 5 \times a \times b,$
$$= 15\,ab.$$

Ex. 2. $\quad 2x \times (-4y^2) = 2(-4)xy^2 = -8xy^2.$

Ex. 3. $\quad \frac{2}{3}a^3 \times 6\,ab^2 \times 11\,b^5 = \frac{2}{3} \times 6 \times 11 \times a^3ab^2b^5 = 44\,a^4b^7.$

These examples illustrate the following method of multiplying two or more monomials:

Multiply the product of the numerical coefficients by the product of the literal factors.

EXERCISES IX.

Multiply·

1.	2.	3.	4.	5.	6.
a	3	$2a$	$-6x$	$5x^2$	$7x^3$
2	x	2	3	-3	-3

7.	8.	9.	10.	11.	12.
$5x$	$-6m$	$3a^2$	$2a$	$-5x^4$	$-2y$
x	m	$-a$	$-3a^3$	$3x$	$-7y^5$

13. $3x^2$ by $5x^2$.   14. $-4x^3$ by $2x^2$.
15. $-5y^3$ by $-2y^4$.   16. $3\,ab$ by $-5\,a^2b^2$.
17. $-4\,x^2y^2$ by $3\,xy$.   18. $-4\,m^3n^3$ by $-5\,m^2n^4$.
19. $5\,abc^2$ by $-3\,abc^3$.   20. $6\,a^2bc^3$ by $-3\,ab^2c^2$.

Find the products of:

21. $3\,x^2y,\ -4\,xy^2,$ and $2\,x^3y.$ 22. $4\,ab^3,\ -2\,a^2b^2,$ and $-a^3b.$
23. $2\,a^3b^2c,\ -2\,ab^3c^2,\ a^4bc^4,$ and $-5\,a^2b^2c^3.$

Multiplication of a Multinomial by a Monomial.

27. If the indicated operation within the parentheses in the product, $4(2+3-1)$, be first performed, we have
$$4(2+3-1) = 4 \times 4 = 16.$$

But if each term within the parentheses be multiplied by 4 and the resulting products be then added, we have

$$4 \times 2 + 4 \times 3 - 4 \times 1 = 8 + 12 - 4 = 16, \text{ as above.}$$

Therefore $\quad 4(2+3-1) = 4 \times 2 + 4 \times 3 - 4 \times 1.$

This example illustrates the following method of multiplying a multinomial by a monomial:

Multiply each term of the multinomial by the monomial, and add algebraically the resulting products. That is,

$$a(b + c - d) = ab + ac - ad.$$

28. Ex. 1. Multiply $(x - y)$ by 3.

We have $\quad 3(x - y) = 3x - 3y.$

Ex. 2. Multiply $3x - 2y - 7z$ by $-4x.$

We have

$$4x(3x - 2y - 7z) = (-4x)(3x) - (-4x)(2y) - (-4x)(7z)$$
$$= -12x^2 + 8xy + 28xz.$$

Such steps as changing $(-4x)(3x)$ into $-12x^2$, $-(-4x)(2y)$ into $+8xy$, and $-(-4x)(7z)$ into $+28xz$, should be performed mentally.

The work may be arranged as in arithmetic, by placing the multiplier under the multiplicand:

$$\begin{array}{r} 3x - 2y - 7z \\ -4x \\ \hline -12x^2 + 8xy + 28xz \end{array}$$

It is customary to multiply from left to right, instead of from right to left, as in arithmetic.

EXERCISES X.

Multiply

1. $a + 1$ by 2.
2. $x - 1$ by $-2.$
3. $a - 4$ by $3.$
4. $a + b$ by 3.
5. $a - b$ by $-5.$
6. $a - 2b$ by $-6.$
7. $x + y$ by $3x.$
8. $x - y$ by $-2x.$
9. $3x - 2y$ by $-5y.$
10. $x^2 + x + 1$ by $3x.$
11. $a^2 - 2a + 1$ by $-2a.$
12. $3m^2 - 2n + 3$ by $-2m.$
13. $5y^2 - 3y + 2$ by $-3y.$

Multiply $a^3 - 3a^2 + a - 1$ by

14. 5.   **15.** $2a$.   **16.** $-3ab$.   **17.** $5a^2b$.

Multiply $x^2 - 3xy + 2y^2$ by

18. 3.   **19.** $2x$.   **20.** $-3y^2$.   **21.** $-4x^2y$.

Multiplication of Multinomials by Multinomials.

29. Ex. Multiply $7 - 5$ by $2 + 3$.

If we let a stand for $7 - 5$, we have

$$(2+3)a = 2a + 3a.$$

Now replacing a by $7 - 5$, we obtain

$$(2+3)(7-5) = 2(7-5) + 3(7-5)$$
$$= 2 \times 7 - 2 \times 5 + 3 \times 7 - 3 \times 5.$$

This example illustrates the following method of multiplying a multinomial by a multinomial:

Multiply each term of the multiplicand by each term of the multiplier, and add algebraically the resulting products.

In general,

$$(a+b)(c+d-e) = a(c+d-e) + b(c+d-e)$$
$$= ac + ad - ae + bc + bd - be.$$

30. Ex. 1. Multiply $-3a + 2b$ by $2a - 3b$.

We have
$$\begin{array}{r} -3a + 2b \\ 2a - 3b \\ \hline \end{array}$$

$2a(-3a+2b) = -6a^2 + 4ab$
$-3b(-3a+2b) = + 9ab - 6b^2$
$ \overline{-6a^2 + 13ab - 6b^2}$

The work is arranged as follows: *Write the multiplier under the multiplicand; the first partial product*, i.e., *the product of the multiplicand by the first term of the multiplier, under the multiplier; the second partial product under the first; and so on, placing like terms of the partial products in the same column.*

Ex. 2. Multiply $x+a$ by $x+b$.

We have
$$x + a$$
$$x + b$$
$$\overline{x^2 + ax}$$
$$bx + ab$$
$$\overline{x^2 + (a+b)x + ab}$$

Ex. 3. Multiply $x^2 + y^2 + 1 - xy - x - y$ by $x + y + 1$.
Arranging to descending powers of x, we have

$$x^2 - xy - x + y^2 - y + 1$$
$$x + y + 1$$
$$\overline{}$$
$$x^3 - x^2y - x^2 + xy^2 - xy + x$$
$$x^2y - xy^2 - xy + y^3 - y^2 + y$$
$$+ x^2 - xy - x + y^2 - y + 1$$
$$\overline{x^3 - 3xy + y^3 + 1}$$

Zero in Multiplication.

31. Since $\quad N \cdot 0 = N(b - b)$, by definition of 0,
$$= Nb - Nb = 0,$$
we have $\quad\boldsymbol{N} \cdot \boldsymbol{0} = \boldsymbol{0}$ and $\boldsymbol{0} \cdot \boldsymbol{N} = \boldsymbol{0}$.

That is, *a product is 0 if one of its factors be zero.*

EXERCISES XI.

Multiply

1. $x+1$ by $x+2$. 2. $a-1$ by $a+2$. 3. $m+3$ by $m-4$.
4. $y+8$ by $y-6$. 5. $x-9$ by $x+11$. 6. $x-12$ by $x+15$
7. $m+18$ by $m-7$. 8. $2x+1$ by $x+5$.
9. $3x-2$ by $x-6$. 10. $5y-12$ by $3y+8$.
11. $6m+9$ by $8m-11$. 12. $9y-3$ by $12y+3$.
13. $a+b$ by $a-b$. 14. $x+2y$ by $x+3y$.
15. $a+6b$ by $a-3b$. 16. $m+6n$ by $m-3n$.

EQUATIONS AND PROBLEMS.

17. $x - 11y$ by $x + 3y$.
18. $2a + 3b$ by $5a - 4b$.
19. $3x - 5y$ by $2x + 9y$.
20. $5x + 7z$ by $3x - 11z$.
21. $x^2 + x + 1$ by $x + 1$.
22. $x^2 - 3x + 1$ by $x - 2$.
23. $x^2 + 8x - 7$ by $x - 5$.
24. $x^3 + 3x^2 - 5$ by $x^2 + 2$.
25. $x^4 - 5x^2 + 1$ by $x^2 - 5$.
26. $x^5 - 6x^2 + 2$ by $x^2 - 3$.
27. $x^2 + xy + y^2$ by $x - y$.
28. $a^2 - ab + b^2$ by $a + b$.

Perform the indicated multiplications:

29. $(x + y + 1)(x + y - 1)$.
30. $(a + b + 5)(a + b - 3)$.
31. $(1 + 2x + 2x^2)(1 - 2x + 2x^2)$.
32. $(1 + 4a + 5a^2)(1 - 4a + 5a^2)$.
33. $(1 + ab + a^2b^2)(1 - ab + a^2b^2)$.
34. $(2x^2y^2 + 3xy - 5)(2x^2y^2 - 3xy - 5)$.
35. $(x^2 + xy^2 + y^4)(x^2 - xy^2 + y^4)$.
36. $(a^4 + 10a^2b^2 + 11b^4)(a^4 - 10a^2b^2 + 11b^4)$.

EQUATIONS AND PROBLEMS.

32. Ex. Find the value of x from the equation
$$3(x - 4) + 5 = 4(x - 3).$$

Removing parentheses, $3x - 12 + 5 = 4x - 12.$

Cancelling -12, $\qquad 3x + 5 = 4x.$

Transferring terms, $\qquad 3x - 4x = -5,$

or $\qquad\qquad\qquad\qquad -x = -5.$

Dividing by -1, $\qquad\quad x = 5.$

Check: $3(5-4) + 5 = 4(5-3)$, or $3 + 5 = 4 \times 2$, or $8 = 8$.

To solve such equations: *Remove parentheses, and proceed as in Art. 21.*

Pr. A number of persons were to raise a fund by paying $5 each. Had there been 4 persons more, each would have had to contribute only $3. How many persons were there?

Let x stand for the number of persons.

Then the number of dollars contributed was $5x$.

Had there been 4 persons more, there would have been $x + 4$ persons.

Then the number of dollars contributed would have been $3(x + 4)$.

The problem implies,

in *verbal* language: *The number of dollars contributed in the one case is equal to the number of dollars contributed in the other;*

in *algebraic* language: $\quad 5x = 3(x + 4)$.

Removing parentheses, $5x = 3x + 12$.

Transferring $3x$, $\quad\quad 2x = 12$.

Dividing by 2, $\quad\quad\quad x = 6$.

Check: 6 persons contributed $6 \times 5, = 30$ dollars; $6 + 4$, or 10, persons would have contributed $10 \times 3, = 30$ dollars.

EXERCISES XII.

Solve the following equations:

1. $2(x + 1) = 3x$.
2. $3(x - 2) = x$.
3. $5(2x + 9) = 19x$.
4. $6(1 - 3x) = -12x$.
5. $(x + 1) + 2(x + 2) = 14$.
6. $(x - 2) + 3(x - 3) = -19$.
7. $5(x + 3) - 3(x + 5) = 4$.
8. $7(x - 3) - 3(x - 7) = -12$.
9. $x + 1 - 2(x + 2) = 3(x - 5)$.
10. $2(x - 1) - 3(x - 2) = 4(x - 4)$.
11. $5(x + 2) + 6(x - 2) = 5(x + 8)$.
12. $4(x - 7) - 5(x - 6) = -2(x - 6)$.

13. If 2 is added to a number, and the sum is multiplied by 3, the product will exceed the number by 24. What is the number?

14. If 25 is subtracted from a number, and the remainder is multiplied by 2, the number will exceed this remainder by 10. What is the number?

15. The sum of two numbers is 80. If four times the less exceeds twice the greater by 20, what are the numbers?

16. In an orchard there are 98 pear trees and apple trees. If five times the number of apple trees exceeds three times the number of pear trees by 282, how many trees of each kind are in the orchard?

17. A is x years old, and B twice as old. How old is B? How old was B 16 years ago? What was four times B's age 16 years ago?

18. A father is twice as old as his son; 16 years ago he was four times as old. What are the ages of father and son?

19. Two boys, A and B, had the same number of apples. A said to B: "Give me 18 apples and I will have twice as many as you will have left." How many apples had each?

20. Add 12 to a certain number, and multiply the sum by 8; or subtract 8 from the same number, and multiply the difference by 12. The results will be equal. What is the number?

21. A is 54 years old, and B is 10 years old. After how many years will A be three times as old as B?

22. A father is 32 years older than his son; 10 years ago he was three times as old. What are the ages of father and son?

23. A and B invested equal amounts. A gained $400, and B gained $2000. If B then had twice as much invested as A, how much did each invest?

24. Three boys, A, B, and C, catch 54 fish. If B catches 6 more fish than A, and C catches twice as many as A and B together, how many fish does each boy catch?

25. A woman bought a certain number of yards of silk at $3 a yard, and 10 yards more of cloth than of silk at $2 a yard. If she paid equal amounts for the silk and the cloth, how many yards of each did she buy?

26. One cask contains 65 gallons, and another 35 gallons. If three times as many gallons are drawn from the larger as from the smaller, the contents of the smaller will be equal to twice the contents of the larger. How many gallons were drawn from each cask?

27. A man has $25 in one-dollar bills and two-dollar bills. If he has 20 bills altogether, how many of each kind has he?

28. A rides his bicycle 10 miles an hour, and B his 8 miles an hour. A rides a certain number of hours, and B rides 4 hours longer. If they ride the same distance, how many hours does each ride? How many miles do they ride?

29. Ten men were to raise a certain fund by contributing equal amounts. But 3 men failed to contribute, and in consequence each of the remaining 7 men had to contribute $6 more. What was to be the original contribution of each? What was the amount of the fund?

30. A man paid $720 for a horse, carriage, and harness. The horse cost five times as much as the harness, and the carriage twice as much as the horse and harness together. How much did he pay for each?

31. Three newsboys together sell 140 papers. The second sells 5 more than the first, and the third sells three times as many as the first and second together. How many papers does each boy sell?

32. A boy, being asked his age, replied: "Multiply the number of years in my age by 2 and subtract 4 from the product, multiply this remainder by 2 and subtract 4 from the product, multiply this remainder by 2 and subtract 4 from the product. This remainder is four times the number of years in my age." How old was he?

DIVISION.

33. One power is said to be *higher* or *lower* than another according as its exponent is *greater* or *less* than the exponent of the other. Thus, a^4 is a higher power than a^3 or b^2, but is a lower power than a^6 or b^7

Quotient of Powers of One and the Same Base.

34. Ex. $\quad a^7 \div a^3 = (aaaaaaa) \div (aaa)$
$= (aaaa) \times (aaa) \div (aaa) = aaaa = a^4 = a^{7-3}$

This example illustrates the following method of dividing a higher power by a lower power of the same base:

The exponent of the quotient is the exponent of the dividend minus the exponent of the divisor; or, stated symbolically,

$$a^m \div a^n = a^{m-n}.$$

We also have

$$a^m \div a^n = 1, \text{ when } m = n.$$

EXERCISES XIII.

Express each of the following quotients as a single power:

1. $2^3 \div 2$.   2. $3^5 \div 3^2$.   3. $x^3 \div x^2$.   4. $a^6 \div a^4$.
5. $x^7 \div x^3$.   6. $a^{16} \div a^9$.   7. $(-a)^6 \div a^5$.   8. $(3x)^5 \div (3x)$.

Division of Monomials by Monomials.

35. Ex. 1. $12a \div 4 = (12 \div 4) \times a, = 3 \times a = 3a$.

Ex. 2. $-27 x^7 \div 3 x^2 = (-27 \div 3) \times (x^7 \div x^2) = -9 \times x^5 = -9 x^5$.

Ex. 3. $15 a^3 b^2 \div (-5 ab^2) = [15 \div (-5)] \times (a^3 \div a) \times (b^2 \div b^2)$
$= -3 a^2.$

These examples illustrate the following method of dividing a monomial by a monomial:

Multiply the quotient of the numerical coefficients by the quotient of the literal factors.

EXERCISES XIV.

Divide

1. $2 \overline{)4x}$ 2. $4 \overline{)-8a}$ 3. $-3 \overline{)-12m}$ 4. $-3a \overline{)12a}$ 5. $-5x \overline{)-15x}$

6. a^2 by a.   7. $-6x^3$ by $2x^2$.   8. $20m^4$ by $-4m^2$.

9. $12m^5$ by $3m^3$.   10. $-18x^5$ by $9x^2$.   11. $-25x^6$ by $-5x^4$.

12. a^2b by $-ab$.   13. $-12a^2b^2$ by $-4a^2b$.

14. $-16a^4b^3$ by $-8a^2b^3$.   15. $9a^5b^2$ by $-3a^2b$.

16. $36x^2y^2z$ by $12xyz$.   17. $18m^5n^4p^2$ by $3m^3n^2p$.

18. $10(a+b)^2$ by $5(a+b)$.   19. $6(x^2+y^2)^3$ by $-3(x^2+y^2)$.

Division of a Multinomial by a Monomial.

36. If the indicated operation within the parentheses in the quotient $(8 + 6 - 4) \div 2$ be first performed, we have

$$(8 + 6 - 4) \div 2 = 10 \div 2 = 5.$$

But if each term within the parentheses be first divided by 2, and the resulting quotients be then added, we have

$$8 \div 2 + 6 \div 2 - 4 \div 2 = 4 + 3 - 2 = 5, \text{ as above.}$$

Therefore $(8 + 6 - 4) \div 2 = 8 \div 2 + 6 \div 2 - 4 \div 2.$

This example illustrates the following method of dividing a multinomial by a monomial:

Divide each term of the multinomial by the monomial, and add algebraically the resulting quotients.

That is,

$$(b + c - d) \div a = b \div a + c \div a - d \div a.$$

37. Ex. 1. Divide $6x^2 - 12x$ by $3x$.

We have $(6x^2 - 12x) \div 3x = 6x^2 \div 3x - 12x \div 3x = 2x - 4.$

EXERCISES XV.

Divide
1. $2a + 2$ by 2.
2. $3a^2 - 6$ by -3.
3. $5x - 15$ by -5.
4. $a^2 + a$ by a.
5. $3x^2 - 6x$ by $-3x$.
6. $4m^2 - 8m$ by $4m$.
7. $14a^2 - 7ab$ by $7a$.
8. $10xy^2 - 15x^2y$ by $5xy$.
9. $4a^2b^2 - 6ab^2$ by $-2ab^2$.
10. $3a^3b^2 - 6a^2b^2 + 12a^2b$ by $3a^2b$.
11. $5m^2n^2 - 10m^2n^3 - 15m^2n$ by $-5m^2n$.
12. $6x^2y^2 - 12x^3y^3 + 18x^4y^4$ by $6x^2y^2$.

Divide $12a^3b^2 - 18a^2b^3 - 24a^3b^3$ by
13. -6.
14. $-3a^2$.
15. $2b^2$.
16. $-2ab$.
17. a^2b^2.

Divide $10x^4y^5 - 20x^3y^4 - 30x^2y^3$ by
18. 10.
19. $-2x$.
20. $5xy$.
21. $-2x^2y$.
22. xy^2.

Zero in Division.

38. Since $0 \div N = (a - a) \div N$, by definition of 0,
$$= a \div N - a \div N = 0.$$

We have $0 \div N = 0$, when N is not equal to 0.

Observe that this relation is proved only when N is not equal to 0.

Division of a Multinomial by a Multinomial.

39. The division of one multinomial by another is performed in a way similar to that of dividing one number by another in Arithmetic.

Ex. Divide 125 by 5.

We have

$$
\begin{array}{r}
125\ \underline{|\ 5} \\
20 \times 5 = 100\ \ \ 20 + 5, = 25 \\
\overline{} \\
125 \quad 20 \times 5 = \ 25 \\
25
\end{array}
$$

The work is equivalent to the following:

$$125 \div 5 = 20 + (125 - 20 \times 5) \div 5 = 20 + 25 \div 5 = 25.$$

40. The number 20, obtained by the first step of the division, is called the **Partial Quotient** at that stage. It is the greatest number whose product by the divisor is equal to or less than the dividend.

The above example illustrates the following method of dividing one number by another:

Obtain the partial quotient by inspection.

Subtract the partial quotient times the divisor from the dividend, and divide the remainder by the divisor. Add this quotient to the partial quotient previously obtained.

If D be the given *dividend*, d the given divisor, and q the *partial quotient*, the principle enunciated above, stated symbolically, is:

$$D \div d = q + (D - qd) \div d.$$

41. The following example illustrates the application of this principle in dividing one multinomial by another.

Ex. Divide $x^2 + 3x + 2$ by $x + 1$.

We have

$$(x^2+3x+2) \div (x+1) = x + [(x^2+3x+2) - x(x+1)] \div (x+1) \quad (1)$$
$$= x + (x^2+3x+2-x^2-x) \div (x+1) \quad (2)$$
$$= x + (2x+2) \div (x+1) \quad (3)$$
$$= x + 2 + [(2x+2) - 2(x+1)] \div (x+1) \quad (4)$$
$$= x + 2 + 0 \div (x+1)$$
$$= x + 2, \text{ since } 0 \div (x+1) = 0.$$

We take the quotient of the term containing the highest power of x in the dividend by the term containing the highest power of x in the divisor as the partial quotient at each step.

The work may be arranged more conveniently thus:

$$
\begin{array}{l|l}
x^2 + 3x + 2 & \underline{x + 1} \\
& x + 2, \text{ quotient.} \\
\underline{x^2 + x} & \cdots x(x+1) \text{ to be subtracted from } x^2 + 3x + 2 \text{; see (1)} \\
& \text{ and (2) above.} \\
2x + 2 & \cdots \text{ Remainder to be divided by } x + 1 \text{; see (3) above.} \\
\underline{2x + 2} & \cdots 2(x+1) \text{ to be subtracted from } 2x + 2 \text{; see (4).}
\end{array}
$$

42. The method of applying the principle of Art. 40 to the division of multinomials, as illustrated by this example, may be stated as follows:

Arrange the dividend and divisor to ascending or descending powers of some common letter, the letter of arrangement.

Divide the first term of the dividend by the first term of the divisor, and write the result as the first term of the quotient.

Multiply the divisor by this first term of the quotient, and subtract the resulting product from the dividend.

Divide the first term of the remainder by the first term of the divisor, and write the result as the second term of the quotient.

Multiply the divisor by this second term of the quotient, and subtract the product from the remainder previously obtained. Proceed with the second remainder and all subsequent remainders, in like manner, until a remainder zero is obtained, or until the highest power of the letter of arrangement in the remainder is less than the highest power of that letter in the divisor.

In the first case the division is exact; in the second case the quotient at this stage of the work is called the quotient of the division, and the remainder the remainder of the division.

43. Ex. 1. Divide $x^2 - 4x - 5$ by $x - 5$. We have

$$
\begin{array}{l|l}
x^2 - 4x - 5 & x - 5 \\
\underline{x^2 - 5x} & \overline{x + 1} \\
x - 5 & \\
\underline{x - 5} &
\end{array}
$$

Ex. 2. Divide

$a^3b - 15b^4 + 19ab^3 + a^4 - 8a^2b^2$ by $a^2 - 5b^2 + 3ab$.

Arranging dividend and divisor to descending powers of a, we have

$$\begin{array}{l|l} a^4 + a^3b - 8a^2b^2 + 19ab^3 - 15b^4 & a^2 + 3ab - 5b^2 \\ a^4 + 3a^3b - 5a^2b^2 & \overline{a^2 - 2ab + 3b^2} \\ \hline -2a^3b - 3a^2b^2 + 19ab^3 \\ -2a^3b - 6a^2b^2 + 10ab^3 \\ \hline 3a^2b^2 + 9ab^3 - 15b^4 \\ 3a^2b^2 + 9ab^3 - 15b^4 \end{array}$$

Ex. 3. Divide $8x^3 - y^3$ by $2xy + 4x^2 + y^2$.

Arranging the divisor to descending powers of x, we have:

$$\begin{array}{l|l} 8x^3 - y^3 & 4x^2 + 2xy + y^2 \\ 8x^3 + 4x^2y + 2xy^2 & \overline{2x - y} \\ \hline -4x^2y - 2xy^2 - y^3 \\ -4x^2y - 2xy^2 - y^3 \end{array}$$

Observe that the remainder after the first partial division is arranged to descending powers of x.

EXERCISES XVI.

Find the value of the following indicated divisions:

1. $(x^2 + 2x + 1) \div (x + 1)$.
2. $(x^2 - 3x + 2) \div (x - 2)$.
3. $(x^2 + 4x + 3) \div (x + 3)$.
4. $(x^2 + 5x + 6) \div (x + 2)$.
5. $(x^2 + 4x - 5) \div (x - 1)$.
6. $(x^2 - 7x + 10) \div (x - 2)$.
7. $(x^2 - 10x + 21) \div (x - 3)$.
8. $(x^2 - 2x - 48) \div (x - 8)$.
9. $(x^2 - 11x - 60) \div (x + 4)$.
10. $(x^2 - 15x + 36) \div (x - 12)$.
11. $(10x^2 + 3x - 18) \div (2x + 3)$.
12. $(12x^2 - 17x - 5) \div (3x - 5)$.
13. $(x^3 + x^2 - 3x + 9) \div (x + 3)$.

DIVISION.

14. $(x^3 - 5x^2 + 3x + 9) \div (x - 3)$.
15. $(x^3 - 24x + 5) \div (x + 5)$.
16. $(x^3 - 39x + 18) \div (x - 6)$.
17. $(x^3 - 48x + 7) \div (x + 7)$.
18. $(x^3 - 134x + 143) \div (x - 11)$.
19. $(2x^3 - 3x^2 + 11x - 5) \div (2x - 1)$.
20. $(9x^3 + 30x^2 + 7x - 30) \div (3x + 5)$.
21. $(28x^3 - 45x^2 + 25x - 56) \div (4x - 7)$.
22. $(25x^3 + x + 6) \div (5x + 3)$.
23. $(49x^3 - 29x - 6) \div (7x - 6)$.
24. $(x^2 + 2xy + y^2) \div (x + y)$.
25. $(x^2 - 2xy + y^2) \div (x - y)$.
26. $(x^2 + xy - 2y^2) \div (x + 2y)$.
27. $(a^2 - 2ab - 3b^2) \div (a - 3b)$.
28. $(2x^2 + xy - 3y^2) \div (2x + 3y)$.
29. $(5m^2 + 9mn - 18n^2) \div (5m - 6n)$.
30. $(a^3 + 3a^2b + 3ab^2 + b^3) \div (a + b)$.
31. $(x^3 - 3x^2y + 3xy^2 - y^3) \div (x - y)$.
32. $(12a^3 + 2a^2b - 24ab^2 - 9b^3) \div (2a - 3b)$.
33. $(27m^3 - 36m^2 + 15m - 2) \div (3m - 2)$.
34. $(14a^3 + 47a^2b - 25ab^2 - 63b^3) \div (2a + 7b)$.
35. $(15x^3 - 31x^2y + 19xy^2 - 7y^3) \div (5x - 7y)$.
36. $(a^3 + 1) \div (a + 1)$.   37. $(x^3 + 8) \div (x^2 - 2x + 4)$.
38. $(m^3 - 1) \div (m - 1)$.   39. $(a^3 - b^3c^3) \div (a - bc)$.
40. $(a^2 + b^2 + c^2 + 2ab + 2ac + 2bc) \div (a + b + c)$.
41. $(x^2 - y^2 - 2yz - z^2) \div (x + y + z)$.
42. $(a^2 - b^2 - c^2 + 2bc) \div (a - b + c)$.
43. $(a^4 + a^2 + 1) \div (a^2 - a + 1)$.
44. $(a^5 - 1) \div (a - 1)$.   45. $(a^6 + 1) \div (a^2 + 1)$.
46. $(a^8 - 1) \div (a^3 + a - a^2 - 1)$.   47. $(x^9 + 1) \div (x^2 - x + 1)$.

CHAPTER IV.

INTEGRAL ALGEBRAIC EQUATIONS.

We will now distinguish between two kinds of equations.

Identical Equations.

1. An example of the one kind is:
$$(a + b)(a - b) = a^2 - b^2.$$

The first member is reduced to the second member by performing the indicated multiplication.

2. Such an equation is called an **Identical Equation**, or more simply, an **Identity**.

3. Notice that identical equations are true for all values that may be substituted for the literal numbers involved.

E.g., if $a = 5$ and $b = 3$, the above equation becomes
$$8 \times 2 = 25 - 9, \text{ or } 16 = 16.$$

Conditional Equations.

4. An example of the second kind is:
$$x + 1 = 3.$$

The first member reduces to the second member, when $x = 2$. It seems evident, and it is proved in School Algebra, Ch. IV., that $x + 1$ reduces to 3 *only* when $x = 2$.

5. Such equations *impose conditions* upon the values of the literal numbers involved. Thus, the equation in Art. 4 imposes the condition that if 1 be added to the value of x, the sum will be 3.

Ch. IV, 1–11] INTEGRAL ALGEBRAIC EQUATIONS. 69

A **Conditional Equation** is an equation one of whose members can be reduced to the other only for certain definite values of one or more letters contained in it.

Whenever the word *equation* is used in subsequent work we shall understand by it a *conditional equation*, unless the contrary is expressly stated.

6. An **Integral Algebraic Equation** is an equation whose members are integral algebraic expressions in an unknown number or unknown numbers.

E.g., $3x^2 - 4 = 2x$, and $\frac{2}{3}x + 5y = \frac{1}{4}$ are integral equations.

7. The **Degree** of an integral equation is the degree of its term of highest degree in the unknown number or numbers.

8. A **Linear** or **Simple Equation** is an equation of the *first* degree.

E.g., $x + 1 = 6$ is a linear equation in one unknown number.

9. A **Solution** of an equation is a value of the unknown number, or a set of values of the unknown numbers, which, if substituted in the equation, converts it into an identity.

E.g., 2 is a solution of the equation $x + 1 = 3$,
since, when substituted for x in the equation, it converts the equation into the identity $2 + 1 = 3$.

The set of values 1 and 2, of x and y, respectively, is a solution of the equation $x + y = 3$, since $1 + 2 = 3$ is an identity.

10. To **Solve** an equation is to find its solution.

An equation is said *to be satisfied by its solution*, or *the solution is said to satisfy the equation*, since it converts the equation into an identity.

11. When the equation contains only one unknown number, a solution is frequently called a **Root** of the equation.

E.g., 2 is a root of the equation $x + 1 = 3$.

Equivalent Equations.

12. Consider the solution of the equation

$$\tfrac{3}{4}x - 5 = 1. \tag{1}$$

Adding 5 to both members,

$$\tfrac{3}{4}x = 6. \tag{2}$$

Dividing by 3, $\quad\tfrac{1}{4}x = 2.\tag{3}$

Multiplying by 4, $\quad x = 8.\tag{4}$

It is evident that 8 is a root of equations (1), (2), (3), and (4).

In thus applying the principles of Ch. I., Art. 16, we replace the given equation by a simpler one, which has the same root, this equation by a still simpler one, which again has the same root, and so on.

Such equations as (1), (2), (3), and (4) are called **Equivalent Equations.**

In general, *two equations are equivalent when every solution of the first is a solution of the second, and every solution of the second is a solution of the first.*

13. It is important to notice that the use of the principles given in Ch. I., Art. 16, may lead to incorrect results.

Thus, by (iii.), we should be permitted to multiply both members of an equation by an expression which contains the unknown number.

E.g., the equation $x - 3 = 0$ has the root 3.

Multiplying both members by $x - 2$, we obtain

$$(x - 3)(x - 2) = 0.$$

This equation has the root 3,

since $\quad (3 - 3)(3 - 2) = 0 \cdot 1 = 0;$

and also the root 2,

since $\quad (2 - 3)(2 - 2) = -1 \cdot 0 = 0.$

But 2 is not a root of the given equation, since $2-3$ does not equal 0.

That is, *in multiplying both members by $x-2$, we gained a root 2.* Observe that this root is the root of $x-2=0$.

The derived equation is therefore not equivalent to the given one.

Again, by (iii.), we should be permitted to multiply both members of an equation by 0.

Multiplying both members of $x-3=0$, by 0, we have
$$0(x-3)=0.$$
Any number is a root of this equation, since
$$0(1-3)=0,\ 0(2-3)=0,\ 0(3-3)=0,\ 0(4-3)=0,\ \text{etc.}$$

Finally, by (iv.), we should be permitted to divide both members of any equation by an expression which contains the unknown number.

E.g., the equation $(x-1)(x+1) = 3(x-1)$,

has the root 1, since
$$(1-1)(1+1) = 3(1-1),\ \text{or}\ 0\times 1 = 3\times 0,\ \text{or}\ 0=0;$$
and the root 2, since
$$(2-1)(2+1) = 3(2-1),\ \text{or}\ 1\times 3 = 3\times 1.$$

Dividing both members by $x-1$, we obtain
$$x+1=3.$$

This equation has the root 2 *only*, and not the root 1 of the given equation.

That is, *in dividing both members by $x-1$, we lost the root 1.* Observe that this root is a root of $x-1=0$.

The derived equation is therefore not equivalent to the given one.

14. The correct statements of the principles which are applied in solving equations are, therefore, as follows:

(i.) **Addition and Subtraction.** — *The equation obtained by adding to, or subtracting from, both members of an equation the same number or expression is equivalent to the given one.*

(ii.) **Multiplication and Division.** — *The equation obtained by multiplying or dividing both members of an equation by the same number, not 0, or by an expression which does not contain the unknown number or numbers, is equivalent to the given one.*

These principles are proved in School Algebra, Ch. IV.

In the solutions of equations in the preceding chapters, we multiplied or divided only by Arabic numerals. Nevertheless, we required each result to be checked.

EXERCISES I.

Solve each of the following equations:

1. $x(x+1) = x(x+2)$.
2. $2x(x-3) = 2x(x+4)$.
3. $4(x-1)(x+1) = 4(x-1)(x+2)$.
4. $3(2-x)(x-2) = 3(3-x)(x-2)$.
5. $(2x+7)(x-3) = (x-3)(2x+8)$.
6. $(x+1)(x+2) = (x-3)(x-4)$.
7. $(5x-2)(3x-4) = (3x-5)(5x+6)$.
8. $x(x+2) + x(x+1) = 2x(x+3)$.
9. $x(x-1) + x(2x-1) = (3x-7)(x+2)$.
10. $6x - [7x - (8x-18)] = 17$
11. $x^2 - x[1 - x - 2(3-x)] = x+1$.

Problems.

15. Pr. 1. A carriage, starting from a point A, travels 35 miles daily; a second carriage, starting from a point B, 84 miles behind A, travels in the same direction 49 miles daily. After how many days will the second carriage overtake the first? At what distance from B will the meeting take place?

Let x stand for the number of days after which they meet. Then the number of miles travelled by the first carriage is $35x$, and the number of miles travelled by the second carriage is $49x$.

14-15]   INTEGRAL ALGEBRAIC EQUATIONS.   73

The problem states,

in *verbal* language: *The number of miles travelled by the first carriage is equal to the number of miles travelled by the second carriage minus* 84;

in *algebraic* language: $35x = 49x - 84$.

Transferring $49x$, $14x = -84$.

Dividing by -14, $x = 6$,

the number of days after which the second carriage overtakes the first.

The distance travelled by the first carriage is 210 miles, and the distance travelled by the second carriage is 294 miles. They therefore meet 294 miles from B.

Pr. 2. A man has $4.50 in dimes and dollars, and he has five times as many dimes as dollars. How many coins of each kind has he?

Let x stand for the number of dollars.

Then $5x$ stands for the number of dimes.

We must first express the dimes as fractional parts of dollars, or the dollars as multiples of dimes. The latter method is the simpler. Since one dollar is 10 dimes. x dollars are $10x$ dimes.

The man evidently has 45 dimes.

The problem states,

in *verbal* language: *Ten times the number of dollars plus the number of dimes is equal to* 45;

in *algebraic* language: $10x + 5x = 45$,

$$15x = 45;$$

whence $x = 3$,

the number of dollars.

Then $5x, = 15$, the number of dimes.

Evidently the value of the coins is $3 + \frac{15}{10}$ dollars, or $4.50.

As in this problem, the magnitudes of all concrete quantities of the same kind must be referred to the same unit; if x stand for a certain number of yards, then all other distances must likewise stand for numbers of yards, not of miles or of feet.

EXERCISES II.

1. The sum of three consecutive numbers exceeds the second by 18. What are the numbers?

2. Since 2×1, 2×2, 2×3, etc., are even numbers, $2x$ represents *any even* number. What is the even number next greater than $2x$? Than $2x+2$?

3. From $2 \times 1 + 1$, $2 \times 2 + 1$, $2 \times 3 + 1$, etc., express *any odd* number.

4. The sum of three consecutive even numbers is 72. What are the numbers?

5. The sum of four consecutive odd numbers is 48. What are the numbers?

6. A and B divide a sum of money. A receives $2 as often as B receives $3. If A receives $2 x, how many dollars does B receive?

7. A and B divide $800. A receives $2 as often as B receives $3. How many dollars does each receive?

8. The length of a room is three times its width. If it were 10 feet shorter and 10 feet wider, it would be square. What are the dimensions of the room?

9. A man travels 200 miles by train, boat, and stage. He travels 30 miles farther by boat than by stage, and four times as far by train as by boat and stage together. How many miles does he travel by each conveyance?

10. A man paid a debt in four monthly payments. He paid $50 more each month than the preceding. If his debt was three times his last payment, how much was his first payment? How much was his debt?

11. In a number of two digits, the units' digit is twice the tens' digit. The number itself exceeds twice the units' digit by 24. What is the number?

12. In a number of two digits, the units' digit is twice the tens' digit. If the digits are interchanged, the resulting number lacks 12 of twice the original number. What is the number?

13. Three boys, A, B, and C, have a number of marbles. A and B have 44, B and C have 43, and A and C have 39. How many marbles has each boy?

14. A man, wishing to give alms to several beggars, lacks 10 cents of enough to give 15 cents to each one. If he were to give 12 cents to each one, he would have 14 cents left over. How many beggars are there?

15. A, travelling 20 miles a day, has 4 days' start of B, who travels 25 miles a day. After how many days will B overtake A?

16. A sum of money is equally divided among four persons. If $60 more be divided equally among six persons, the shares will be the same as before. How many dollars are divided?

17. Atmospheric air is a mixture of four parts of nitrogen with one of oxygen. How many cubic feet of oxygen are there in a room 10 yards long, 5 yards wide, and 12 feet high?

18. A merchant paid $6.15 in an equal number of dimes and five-cent pieces. How many coins of each kind did he pay?

19. A man has $4.75 in dimes and quarters, and he has 5 more quarters than dimes. How many coins of each kind has he?

20. The circumference of the front and hind wheels of a wagon are 2 and 3 yards, respectively. What distance has the wagon moved when the front wheel has made 10 revolutions more than the hind wheel?

21. One barrel contained 48 gallons, and another 88 quarts of wine. From the first twice as much wine was drawn as from the second; the first then contained three times as much wine as the second. How much wine was drawn from each?

22. A regiment moves from A to B, marching 20 miles a day. Two days later a second regiment leaves B for A, and marches 30 miles a day. At what distance from A do the regiments meet, A being 340 miles from B?

TYPE-FORMS.

1. We shall in this chapter consider a number of products and quotients which are of frequent occurrence. They enable us to shorten work by writing similar products and quotients without performing the actual multiplications and divisions. They are called **Type-Forms**.

TYPE-FORMS IN MULTIPLICATION.

The Square of a Binomial.

2. By actual multiplication, we have

$$(a+b)^2 = (a+b)(a+b) = a^2 + ab + ba + b^2 = a^2 + 2ab + b^2.$$

That is, *the square of the sum of two numbers is equal to the square of the first number, plus twice the product of the two numbers, plus the square of the second number.*

E.g.,
$$(2x+5y)^2 = (2x)^2 + 2(2x)(5y) + (5y)^2$$
$$= 4x^2 + 20xy + 25y^2.$$

3. By actual multiplication, we have

$$(a-b)^2 = (a-b)(a-b) = a^2 - ab - ba + b^2 = a^2 - 2ab + b^2.$$

That is, *the square of the difference of two numbers is equal to the square of the first number, minus twice the product of the two numbers, plus the square of the second number.*

E.g.,
$$(3x-7y)^2 = (3x)^2 - 2(3x)(7y) + (7y)^2$$
$$= 9x^2 - 42xy + 49y^2.$$

4. We have $(5\,a^3)^2 = 5 \cdot 5\, a^3 a^3 = 5^2\, a^{3+3} = 5^2\, a^{2\times 3} = 25\, a^6.$

That is, to square a monomial: *Square the numerical coefficient, and multiply the exponent of each literal factor by 2.*

EXERCISES I.

Write, without performing the actual multiplications, the values of

1. $(a+1)^2$
2. $(a-2)^2$
3. $(a+5)^2.$
4. $(a-7)^2$
5. $(m-10)^2.$
6. $(2x+1)^2.$
7. $(3x-2)^2.$
8. $(5a-6)^2.$
9. $(10a-3)^2.$
10. $(m+n)^2.$
11. $(m-p)^2.$
12. $(2x-y)^2.$
13. $(5a+b).$
14. $(6m-n)^2.$
15. $(2a+3b)^2.$
16. $(3a-5b)^2.$
17. $(xy+1)^2.$
18. $(mn-2)^2.$
19. $(ab+c)^2.$
20. $(2xy-3z)^2.$
21. $(5mn+2p)^2.$
22. $(x^2+1)^2.$
23. $(x^2-5)^2.$
24. $(2x^2+3)^2.$
25. $(3x^2-5y^2)^2.$
26. $(2a^2b^2+c^2)^2.$
27. $(5x^2y^3-z^3)^2$

Product of the Sum and Difference of Two Numbers.

5. By actual multiplication, we have

$$(a+b)(a-b) = a^2 - ab + ba - b^2 = a^2 - b^2.$$

That is, *the product of the sum of two numbers and the difference of the same numbers, taken in the same order, is equal to the square of the first, minus the square of the second.*

Ex. 1. $(2x+2y)(2x-3y) = (2x)^2 - (3y)^2 = 4x^2 - 9y^2.$

EXERCISES II.

Find, without performing the actual multiplications, the values of

1. $(x+1)(x-1).$
2. $(a-5)(a+5).$
3. $(m+6)(m-6).$
4. $(y-7)(y+7).$
5. $(3a+1)(3a-1).$
6. $(5x-4)(5x+4).$

7. $(2x-7)(2x+7)$. 8. $(3a+10)(3a-10)$.
9. $(5+2a)(5-2a)$. 10. $(7-3b)(7+3b)$.
11. $(9-4x)(9+4x)$. 12. $(x+y)(x-y)$.
13. $(m-n)(m+n)$. 14. $(2a+b)(2a-b)$.
15. $(3x+y)(3x-y)$. 16. $(2x+3y)(2x-3y)$.
17. $(5m-3n)(5m+3n)$. 18. $(ab+1)(ab-1)$.
19. $(4ax+3)(4ax-3)$. 20. $(xy+z)(xy-z)$.
21. $(3ab+5c)(3ab-5c)$. 22. $(x^2-1)(x^2+1)$.
23. $(a^3+1)(a^3-1)$. 24. $(3a^4+2)(3a^4-2)$.
25. $(5a^3+2b)(5a^3-2b)$. 26. $(a+b)(a-b)(a^2+b^2)$.

The Product $(x+a)(x+b)$.

6. By actual multiplication, we have

$$(x+a)(x+b) = x^2 + ax + bx + ab = x^2 + (a+b)x + ab;$$
$$(x+a)(x-b) = x^2 + ax - bx - ab = x^2 + (a-b)x - ab;$$
$$(x-a)(x-b) = x^2 - ax - bx + ab = x^2 - (a+b)x + ab.$$

We thus derive the following method for multiplying two binomials which have a common first term:

The first term of the product is the square of the common first terms of the binomials.

The coefficient of the second term of the product is the algebraic sum of the second terms of the binomials.

The last term of the product is the product of the last terms of the binomials.

Ex. 1. Write the product $(x+3)(x+7)$.

The first term is x^2;

The second term is $(3+7)x$, $=10x$;

The third term is $3 \times 7 = 21$.

Therefore $(x+3)(x+7) = x^2 + 10x + 21$.

TYPE-FORMS IN MULTIPLICATION.

Ex. 2. Write the product $(x-8)(x+2)$.
First term: x^2; second term: $(-8+2)x, =-6x$;
third term: $-8 \times 2 = -16$.
Therefore $(x-8)(x+2) = x^2 - 6x - 16$.

Ex. 3. Write the product $(a^2+9)(a^2-3)$.
First term: $(a^2)^2, = a^4$; second term: $(9-3)a^2, = 6a^2$;
third term: $9 \times (-3), = -27$.
Therefore $(a^2+9)(a^2-3) = a^4 + 6a^2 - 27$.

Ex. 4. Write the product $(x-5y)(x-7y)$.
First term: x^2; second term: $(-5y-7y)x, = -12xy$;
third term: $-5y \times (-7y), = 35y^2$.
Therefore $(x-5y)(x-7y) = x^2 - 12xy + 35y^2$.

EXERCISES III.

Find, without performing the actual multiplications, the values of:

1. $(x+1)(x+2)$.
2. $(x+1)(x-2)$.
3. $(x-1)(x+2)$.
4. $(x-1)(x-2)$.
5. $(x+4)(x+7)$.
6. $(x+4)(x-7)$.
7. $(x-4)(x+7)$.
8. $(x-4)(x-7)$.
9. $(x+2)(x-11)$.
10. $(x-2)(x-11)$.
11. $(x+5)(x+9)$.
12. $(x+5)(x-9)$.
13. $(a+b)(a+2b)$.
14. $(a-b)(a+2b)$.
15. $(a+b)(a-2b)$.
16. $(a-b)(a-2b)$.
17. $(x+3y)(x-5y)$.
18. $(x-3y)(x-5y)$.
19. $(5+3x)(5-x)$.
20. $(3+5a)(5a-2)$.
21. $(9-7x)(9-2x)$.
22. $(ab+5)(ab-3)$.
23. $(ab-7)(ab+2)$.
24. $(ab+c)(ab+2c)$.
25. $(xy-7z)(xy+5z)$.
26. $(a^2b+3c)(a^2b-5c)$.
27. $(x^2+3y^2)(x^2-5y^2)$.
28. $(m^2+3n^2)(m^2-11n^2)$.
29. $(x^2+5ab)(x^2+6ab)$.
30. $(x^2+5ab)(x^2-6ab)$.

TYPE–FORMS IN DIVISION.

Quotient of the Sum or the Difference of Like Powers of Two Numbers by the Sum or the Difference of the Numbers.

7. By actual division, we obtain

$$(a^2 - b^2) \div (a + b) = a - b \text{ and } (a^2 - b^2) \div (a - b) = a + b.$$

That is, *the difference of the squares of two numbers is divisible by the sum, and also by the difference of the numbers. The quotient in the first case is the difference of the numbers, taken in the same order, and in the second case is the sum of the numbers.*

Ex. 1. $(9 - 25x^2) \div (3 + 5x) = 3 - 5x.$

Ex. 2. $(16x^4 - 81y^6) \div (4x^2 - 9y^3) = 4x^2 + 9y^3.$

EXERCISES IV.

Find the values of the following quotients without performing the actual divisions:

1. $(x^2 - 1) \div (x + 1).$
2. $(x^2 - 1) \div (x - 1).$
3. $(a^2 - 4) \div (a + 2).$
4. $(x^2 - 9) \div (x - 3).$
5. $(a^2 - 64) \div (a + 8).$
6. $(m^2 - 100) \div (m + 10).$
7. $(4x^2 - 1) \div (2x + 1).$
8. $(9a^2 - 1) \div (3a - 1).$
9. $(64x^2 - 1) \div (8x - 1).$
10. $(121a^2 - 1) \div (11a + 1).$
11. $(16x^2 - 9) \div (4x + 3).$
12. $(25x^2 - 64) \div (5x - 8).$
13. $(100x^2 - 81) \div (10x + 9).$
14. $(x^2 - y^2) \div (x + y).$
15. $(16x^2 - m^2) \div (4x - m).$
16. $(25a^2 - b^2) \div (5a + b).$
17. $(81m^2 - n^2) \div (9m + n).$
18. $(4a^2 - 9b^2) \div (2a + 3b).$
19. $(16x^4 - 25y^4) \div (4x^2 - 5y^2).$
20. $(100m^4 - 81n^6) \div (10m^2 - 9n^3).$
21. $(a^2b^2 - 1) \div (ab - 1).$
22. $(16a^2b^2 - c^2) \div (4ab + c).$

The Sum and Difference of Two Cubes.

8. By actual division, we obtain

$$(a^3 + b^3) \div (a + b) = a^2 - ab + b^2.$$
$$(a^3 - b^3) \div (a - b) = a^2 + ab + b^2.$$

Ex. 1. $(8 x^3 + \tfrac{1}{125}) \div (2 x + \tfrac{1}{5}) = (2 x)^2 - (2 x)(\tfrac{1}{5}) + (\tfrac{1}{5})^2$
$\qquad = 4 x^2 - \tfrac{2}{5} x + \tfrac{1}{25}.$

Ex. 2. $(a^{12} - b^9) \div (a^4 - b^3) = (a^4)^2 + a^4 b^3 + (b^3)^2$
$\qquad = a^8 + a^4 b^3 + b^6.$

9. We have

$$(4 x^2)^3 = 4 \cdot 4 \cdot 4 \, x^2 x^2 x^2 = 4^3 x^{2+2+2} = 64 \, x^{3 \times 2} = 64 \, x^6.$$

That is, *a given monomial is the cube of a monomial, when its numerical coefficient is a cube and the exponent of each literal factor is divisible by 3.*

EXERCISES V.

Find the values of the following quotients without performing the actual divisions:

1. $(x^3 + 1) \div (x + 1).$
2. $(x^3 - 1) \div (x - 1).$
3. $(a^3 + 8) \div (a + 2).$
4. $(a^3 - 27) \div (a - 3).$
5. $(m^3 - 64) \div (m - 4).$
6. $(125 + r^3) \div (5 + r).$
7. $(1000 - x^3) \div (10 - x).$
8. $(a^3 + x^3) \div (a + x).$
9. $(a^3 - m^3) \div (a - m).$
10. $(8 x^3 + y^3) \div (2 x + y).$
11. $(27 a^3 - b^3) \div (3 a - b).$
12. $(64 m^3 + n^3) \div (4 m + n).$
13. $(8 a^3 - 27 b^3) \div (2 a - 3 b).$
14. $(125 x^3 - 216 y^3) \div (5 x - 6 y).$
15. $(27 a^3 + 64 b^3) \div (3 a + 4 b).$
16. $(8 x^3 + 125 y^3) \div (2 x + 5 y).$
17. $(x^6 + 1) \div (x^2 + 1).$
18. $(a^9 - 1) \div (a^3 - 1).$
19. $(a^{12} + 1) \div (a^4 + 1).$
20. $(a^3 b^3 - 1) \div (ab - 1).$
21. $(8 x^3 y^3 + 27) \div (2 xy + 3).$
22. $(64 x^3 y^3 + z^3) \div (4 xy + z).$
23. $(27 a^3 b^6 - 8 c^9) \div (3 a b^2 - 2 c^3).$

CHAPTER VI.

FACTORS AND MULTIPLES OF INTEGRAL ALGEBRAIC EXPRESSIONS.

INTEGRAL ALGEBRAIC FACTORS.

1. A product of two or more factors is, by the definition of division, exactly divisible by any one of them.

An **Integral Algebraic Factor** of an expression is an integral expression which exactly divides the given one.

E.g., integral factors of $6\,a^2x$ are 6, a^2x, $3\,x$, $2\,a^2$, etc.;

integral factors of $a^2 - b^2$ are $a+b$ and $a-b$.

2. A **Prime Factor** is one which is exactly divisible only by itself and unity.

E.g., the prime factors of $6\,a^2x$ are $2, 3, a, a, x$.

A **Composite Factor** is one which is not prime, *i.e.*, which is itself the product of two or more prime factors.

E.g., composite factors of $6\,a^2x$ are $6, ax, 2\,a, 3\,ax$, etc.

3. Any monomial can be resolved into its prime factors by inspection.

E.g., the prime factors of $4\,a^3b^2$ are $2, 2, a, a, a, b, b$.

Multinomials whose Terms have a Common Factor.

4. From Ch. III., Art. 27, we have

$$ab + ac - ad = a(b+c-d). \qquad (1)$$

This relation may be called the *Fundamental Formula for Factoring.* From it we derive the following method for finding the second factor of a multinomial whose terms have a common factor:

Determine by inspection the remaining factors of its terms, and take their algebraic sum.

5. Ex. 1. Factor $2\,x^2y - 2\,xy^2$.

The factor $2\,xy$ is common to both terms; the remaining factor of the first term is x, that of the second term is $-y$, and their algebraic sum is $x - y$.

Consequently $2\,x^2y - 2\,xy^2 = 2\,xy\,(x - y)$.

Ex. 2. $ab^2 + abc + b^2c = b\,(ab + ac + bc)$.

6. In the fundamental formula the letters a, b, c, d may stand for binomial or multinomial expressions.

Ex. 1. Factor $a\,(x - 2\,y) + b\,(x - 2\,y)$.

The factor $x - 2\,y$ is common to both terms; the remaining factor of the first term is a, that of the second term is b, and their algebraic sum is $a + b$.

Consequently $a\,(x - 2\,y) + b\,(x - 2\,y) = (x - 2\,y)\,(a + b)$.

Ex. 2. Factor $1 - a + x\,(1 - a)$.

We have
$$1 - a + x(1 - a) = 1 \cdot (1 - a) + x(1 - a) = (1 - a)(1 + x).$$

EXERCISES I.

Factor the following expressions:

1. $2\,a + 2$. 2. $3\,x - 6$. 3. $5\,a^2 + 10\,a$. 4. $6\,a^3 - 12\,a$.
5. $14\,m^2 - 21\,m^3$. 6. $6\,x^6 + 8\,x^3$. 7. $4\,x^2 - 6\,xy$.
8. $10\,a^2b + 15\,b^2$. 9. $14\,x^2y - 21\,xy^2$. 10. $x^3y + 6\,xy^3$.
11. $8\,a^3b^2 + 10\,a^2b^3$. 12. $2\,a^2b + 4\,ab^2 - 6\,a^2b^2$.
13. $5\,m^3 - 10\,m^2n + 15\,m^2n^2$. 14. $6\,ab^2c - 8\,a^2bc + 24\,abc^2$.

84   ALGEBRA.   [Ch. VI

15. $15\,x^2y^3z^5 - 20\,xy^3z^3 + 25\,x^2y^2z^2$.
16. $27\,x^2y^3z^2 - 36\,x^2y^2z - 54\,x^2yz^2$.
17. $14\,a^5b^3c - 28\,a^4b^4c^3 - 49\,a^3b^5c^2$.
18. $a(x-y) + b(x-y)$.   **19.** $3x(a+b) - 6y(a+b)$.

Grouping Terms.

7. When all the terms of a given expression do not contain a common factor, it is sometimes possible to group the terms so that all the groups shall contain a common factor.

Ex. 1. Factor $2a + 2b + ax + bx$.

Factoring the first two terms by themselves, and the last two terms by themselves, we obtain:

$$2(a+b) + x(a+b) = (a+b)(2+x).$$

Ex. 2. $\quad x^2 - xy - xz + yz = (x^2 - xy) - (xz - yz)$
$$\qquad\qquad = x(x-y) - z(x-y) = (x-y)(x-z).$$

Ex. 3. $\quad x^3 + 3x^2 - 2x - 6 = (x^3 + 3x^2) - (2x + 6)$
$$= x^2(x+3) - 2(x+3)$$
$$= (x+3)(x^2-2).$$

EXERCISES II.

Factor the following expressions:

1. $a + b + ax + bx$. **2.** $ax - a + bx - b$.
3. $ax - a + b - bx$. **4.** $ax + by + bx + ay$.
5. $1 + 2a + b + 2ab$. **6.** $1 - 3a - b + 3ab$.
7. $x^3 + x^2 + x + 1$. **8.** $x^3 + x - 1 - x^2$.
9. $x^3 + 2 - 2x - x^2$. **10.** $x^3 + x^2 + ax + a$.
11. $x^3 - x^2 - a + ax$. **12.** $a + ab + ab^2 + ab^3$.
13. $x - xy - xy^2 + xy^3$. **14.** $ab - 5b + 3a - 15$.
15. $2mn - 7n + 18m$ 63. **16.** $6ab - 2b + 15a - 5$.
17. $(x+y)^2 + ax + ay$. **18.** $(2a-b)^2 + 4ax - 2bx$.
19. $ax + by + cz + bx + cy + az + cx + ay + bz$.

Use of Type-Forms in Factoring.

8. If an expression is in the form of one of the type-forms considered in Ch. V., or if it can be reduced to such a form, its factors can be written by inspection.

Trinomial Type-Forms.

9. From Ch. V., Arts. 2 and 3, we have:
$$a^2 + 2\,ab + b^2 = (a + b)^2,$$
$$a^2 - 2\,ab + b^2 = (a - b)^2.$$

Therefore a trinomial which is the square of a binomial must satisfy the following conditions:

(i.) *One term of the trinomial is the square of the first term of the binomial.*

(ii.) *A second term of the trinomial is the square of the second term of the binomial.*

(iii.) *The remaining term of the trinomial is twice the product of the two terms of the binomial.*

10. Ex. **1.** Factor $x^2 + 6x + 9$.

x^2 is the square of x, 9 is the square of 3, and $6x = 2 \cdot x \cdot 3$.

Therefore $\qquad x^2 + 6x + 9 = (x + 3)^2$.

Ex. **2.** Factor $-4xy + 4x^2 + y^2$.

$4x^2$ is the square of $2x$, or of $-2x$; y^2 is the square of y, or of $-y$. Since the term $-4xy$ is negative, one term of the binomial is negative, the other positive.

Therefore $-4xy + 4x^2 + y^2 = (2x - y)^2 = (-2x + y)^2$.

EXERCISES III.

Factor the following expressions:

1. $a^2 - 2a + 1$.   2. $x^2 + 4x + 4$.   3. $a^2 + 6a + 9$.
4. $y^2 - 10y + 25$.   5. $m^2 - 12m + 36$.   6. $a^2 + 4ab + 4b^2$.
7. $x^2 - 14xy + 49y^2$.   8. $m^2 - 18m + 81$.   9. $1 + 4z^2 + 4z$.
10. $4 - 4x + x^2$.   11. $x^2y^2 + 2xy + 1$.   12. $m^2n^2 - 4mn + 4$.

13. $9 + 6xy + x^2y^2$.
14. $1 - 12mn + 36m^2n^2$.
15. $x^2 - 8xy + 16y^2$.
16. $25x^2 - 20xz + 4z^2$.
17. $4a^2 + 24ab + 36b^2$.
18. $x^4 - 4x^2 + 4$.
19. $16a^4 - 24a^2b^2 + 9b^4$.
20. $9x^2 + 18x + 9$.
21. $2x^3 - 32x^2 + 128x$.
22. $5a^3 + 20a^2b + 20ab^2$.
23. $(x^2 - 2x)^2 + 2(x^2 - 2x) + 1$.
24. $(x - y)^2 - 2(x - y) + 1$.

11. From Ch. V., Art. 6, we have

$$x^2 + (a + b)x + ab = (x + a)(x + b).$$

When a trinomial, arranged to descending powers of some letter, say x, can be factored into two binomials, it must satisfy the following conditions:

(i.) *One term of the trinomial is the square of the letter of arrangement*, i.e., *of the common first term of the binomial factors.*

(ii.) *The coefficient of the first power of the letter of arrangement in the trinomial is the algebraic sum of two numbers whose product is the remaining term of the trinomial.*

(iii.) *These two numbers are the second terms of the binomial factors.*

12. Ex. 1. Factor $x^2 + 8x + 15$.

The common first term of the binomial factors is evidently x. The second terms are two numbers whose product is 15, and whose sum is 8. By inspection we see that

$$3 + 5 = 8 \text{ and } 3 \times 5 = 15;$$

that is, the second terms of the binomial factors are 3 and 5.

Consequently, $x^2 + 8x + 15 = (x + 3)(x + 5)$.

Ex. 2. Factor $x^2 - 7x + 12$.

The common first term of the binomial factors is x. The second terms are two numbers whose product is 12, and whose sum is -7. Since their product is *positive*, they must be *both positive* or *both negative*; and since their sum is *negative*, they must be *both negative*.

The possible pairs of negative factors of 12 are: -1 and 12; -2 and -6; -3 and -4.

But since $$-3 + (-4) = -7,$$

the second terms of the binomial factors are -3 and -4.

Consequently $x^2 - 7x + 12 = (x-3)(x-4)$.

Ex. 3. Factor $a^2x^2 + 5ax - 24$.

The common first term of the binomial factors is ax. The second terms are two numbers whose product is -24, and whose sum is 5. Since their product is negative, one must be positive and the other negative; and since their sum is positive, the positive number must have the greater absolute value. The possible pairs of factors of -24 are: -1 and 24; -2 and 12; -3 and 8; -4 and 6.

But since $$-3 + 8 = 5,$$

the second terms of the binomial factors are -3 and 8.

Consequently $a^2x^2 + 5ax - 24 = (ax - 3)(ax + 8)$.

Ex. 4. Factor $x^2 - 3xy - 28y^2$.

The common first term of the binomial factors is x. The second terms are two numbers whose product is $-28y^2$, and whose sum is $-3y$. It is evident that both of these terms contain y as a factor. Therefore we have only to find their numerical coefficients.

Since their product is negative, one must be positive and the other negative; and since their sum is negative, the negative number must have the greater absolute value. The possible pairs of factors of -28 are: 1 and -28; 2 and -14; 4 and -7.

But since $$4 + (-7) = -3,$$

the second terms of the binomial factors are $4y$ and $-7y$.

Consequently $x^2 - 3xy - 28y^2 = (x + 4y)(x - 7y)$.

EXERCISES IV.

Factor the following expressions:

1. $x^2 + 3x + 2$.
2. $x^2 - 3x + 2$.
3. $x^2 + x - 2$.
4. $x^2 - x - 2$.
5. $y^2 + 4y + 3$.
6. $y^2 - 4y + 3$.
7. $y^2 - 2y - 3$.
8. $y^2 + 2y - 3$.
9. $m^2 - 5m - 24$.
10. $m^2 + 11m + 24$.
11. $m^2 + 5m - 24$.
12. $m^2 + 14m + 24$.
13. $x^2 - 7x + 6$.
14. $x^2 - 7x + 12$.
15. $x^2 - 7x - 8$.
16. $x^2 - 7x - 18$.
17. $y^2 + 10y - 24$.
18. $y^2 + 23y - 24$.
19. $y^2 - 2y - 24$.
20. $y^2 + 2y - 24$.
21. $y^2 - 14y + 24$.
22. $y^2 - 25y + 24$.
23. $y^2 - 11y + 24$.
24. $y^2 - 10y - 24$.
25. $x^2 - xy - 20y^2$.
26. $x^2 + xy - 20y^2$.
27. $x^2 + 9xy + 20y^2$.
28. $x^2 - 9xy + 20y^2$.
29. $4a^2 + 2ab - 2b^2$.
30. $9m^2 - 36mn - 45n^2$.
31. $4m^2 - 28mn - 72n^2$.
32. $x^2y^2 + xy - 2$.
33. $x^2y^2 - xy - 2$.
34. $16x^2y^2 - 12xy + 2$.
35. $x^2y^2 + xyz - 2z^2$.
36. $x^2y^2 + 3xyz + 2z^2$.
37. $m^2 + 10mnp + 21n^2p^2$.
38. $m^2 + 4mnp - 21n^2p^2$.

Binomial Type-Forms.

13. From Ch. V., Art. 5, we have

$$a^2 - b^2 = (a+b)(a-b).$$

That is, *the difference of the squares of two numbers can be written as the product of the sum and the difference of the numbers.*

Ex. 1.
$$a^2x^2 - \tfrac{1}{4}b^2 = (ax)^2 - (\tfrac{1}{2}b)^2$$
$$= (ax + \tfrac{1}{2}b)(ax - \tfrac{1}{2}b).$$

Ex. 2.
$$32m^4n - 2n^3 = 2n(16m^4 - n^2)$$
$$= 2n[(4m^2)^2 - n^2]$$
$$= 2n(4m^2 + n)(4m^2 - n).$$

Ex. 3. $x^2 - 4xy + 4y^2 - 9z^2 = (x - 2y)^2 - (3z)^2$
$$= (x - 2y + 3z)(x - 2y - 3z).$$

14. The difference of any even powers of two numbers can be written as the difference of the squares of two numbers, and should therefore first be factored by applying this type-form.

Ex. $a^4 - b^4 = (a^2)^2 - (b^2)^2 = (a^2 + b^2)(a^2 - b^2)$
$= (a^2 + b^2)(a + b)(a - b).$

EXERCISES V.

Factor the following expressions:

1. $a^2 - 1$. 2. $x^2 - 4$. 3. $y^2 - 25$. 4. $16 - m^2$.
5. $49 - v^2$. 6. $4x^2 - 1$. 7. $36m^2 - 1$. 8. $81z^2 - 1$.
9. $x^2y^2 - 1$. 10. $9 - m^2n^2$. 11. $16a^2b^2 - 25$. 12. $16a^2 - b^2$.
13. $9x^2 - 4y^2$. 14. $36m^2 - 25n^2$. 15. $100x^2y^2 - 9b^2$.
16. $36a^2b^2 - 49c^2$. 17. $9x^2y^2z^2 - 16$. 18. $x^4 - 1$.
19. $x^8 - y^8$. 20. $a^6 - x^2$. 21. $m^4n^4 - 36p^2$.
22. $1 - 100a^2b^4c^6$. 23. $x^{16} - 625$. 24. $a^{16}b^8 - 100$.

Find, by resolving into factors, the values of:

25. $(183)^2 - (117)^2$. 26. $(553)^2 - (447)^2$.
27. $(273)^2 - (227)^2$. 28. $(543)^2 - (533)^2$.

Factor the following expressions:

29. $(1 + a)^2 - b^2$. 30. $b^2 - (1 + a)^2$. 31. $b^2 - (1 - a)^2$.
32. $(x + y)^2 - 9$. 33. $9 - (x + y)^2$. 34. $9 - (x - y)^2$.
35. $(2a + 1)^2 - 4a^2$. 36. $16m^2 - (1 + 4m)^2$.
37. $(a - b)^2 - c^2$. 38. $c^2 - (a - b)^2$.
39. $(2x + 3y)^2 - 9y^2$. 40. $4x^2 - (2x - 3y)^2$.
41. $x^4 - (x^2 - 1)^2$. 42. $m^6 - (m^3 - 1)^2$. 43. $a^{10} - (a^5 - 1)^2$.
44. $x^4 - 16(x - y)^2$. 45. $a^2b^2 - 9(ab - 1)^2$.
46. $(x + 2)^2 - (x - 4)^2$. 47. $(3 - a)^2 - (3 + b)^2$.
48. $(x + y)^2 - (u + v)^2$. 49. $(m - n)^2 - (p - q)^2$.

15. From Ch. V., Art. 8, we derive:
$$a^3 + b^3 = (a+b)(a^2 - ab + b^2),$$
$$a^3 - b^3 = (a-b)(a^2 + ab + b^2).$$

Ex. 1. $\quad x^3 + 8y^3 = x^3 + (2y)^3$
$$= (x+2y)[x^2 - x(2y) + (2y)^2]$$
$$= (x+2y)(x^2 - 2xy + 4y^2).$$

Ex. 2. $\quad 512\,x^6 - y^3 = (8x^2)^3 - y^3$
$$= (8x^2 - y)[(8x^2)^2 + 8x^2 \times y + y^2]$$
$$= (8x^2 - y)(64x^4 + 8x^2y + y^2).$$

EXERCISES VI.

Factor the following expressions:

1. $a^3 + 1$.
2. $x^3 - 8$.
3. $8a^3 - 1$.
4. $1 - b^3$.
5. $1 + 27y^3$.
6. $1 - 64m^3$.
7. $a^3 + m^3$.
8. $x^3 - z^3$.
9. $125a^3 - b^3$.
10. $x^3 - 27y^3$.
11. $8a^3 + 125b^3$.
12. $343x^3 - 27y^3$.
13. $a^3b^3 + 1$.
14. $x^3y^3 - 1$.
15. $m^3n^3 + p^3$.
16. $a^3b^3 - c^3$.
17. $x^3y^3 - 8u^3$.
18. $27a^3b^3 + c^3$.
19. $x^6 - 64y^3$.
20. $y^6 + 125z^3$.
21. $x^{12} + 343$.
22. $27a^9 - b^6$.
23. $x^9 - y^9$.
24. $x^{12} - y^{12}$.
25. $5x^4 - 40xy^3$.
26. $16a^5 + 250a^2$.
27. $4x^7y - 4xy^7$.

EXERCISES VII.

Factor the following expressions by methods given in this chapter:

1. $x^4 - 2x^2y^2 + y^4$.
2. $x^2 - 12x + 35$.
3. $9 - (3-x)^2$.
4. $a^4 + 2a^3b - 2ab^3 - b^4$.
5. $x^5 - x^3 - x^2 + 1$.
6. $49x^2 - 28xy + 4y^2$.
7. $x(x-1) - x + 1$.
8. $4ax + 2a^2 + 2x^2$.
9. $x^4 - 16x^2 + 63$.
10. $x^2 + (a+b)x + ab$.
11. $16a^2 - 25b^4c^6$.
12. $a^9b^9 + 64c^6$.

13. $1 - 256\, x^8 y^8$.
14. $x^3 - 8 - 6\, x^2 + 12\, x$.
15. $6\, a^2 x^2 - 3\, a^2 x^3 - 3\, a^2 x$.
16. $3\, a(a-1) - 3(a-1)$.
17. $x^2 + (p-q)x - pq$.
18. $n^6 + 5\, n^4 x^2 + 5\, n^2 x^4 + x^6$.
19. $(x^2 + x + 1)^2 - (x^2 - x + 1)^2$.
20. $x^2(x+1) - b^2(b+1)$.

HIGHEST COMMON FACTORS.

16. If two or more integral algebraic expressions have no common factor except 1, they are said to be *prime to one another*.

E.g, ab and cd; $5\, x^2 y$ and $8\, z^3$; $a^2 + b^2$ and $a^2 - b^2$

17. The **Highest Common Factor (H. C. F.)** of two or more integral algebraic expressions is the expression of highest degree which exactly divides each of them.

E.g., the H. C. F. of ax^2, bx^3, and cx^4 is evidently x^2.

18. Monomial Expressions. — The H. C. F. of monomials can be found by inspection.

Ex. 1. Find the H. C. F. of $x^2 y^5 z$, $x^4 y^3 z^3$, and $x^3 y^4 z^4$.

In the expression of highest degree which exactly divides each of the given expressions, the highest power of x is evidently x^2, of y is y^3, and of z is z. Therefore the required H. C. F. is $x^2 y^3 z$.

Observe that the power of each letter in the H. C. F. is the *lowest* power to which it occurs in any of the given expressions.

If the monomials contain numerical factors, the Greatest Common Measure (G. C. M.) of these factors should be found as in Arithmetic.

Ex. 2. Find the H. C. F. of $18\, a^4 b^5 c^3 d$, $42\, a^3 bc^4$, and $30\, a^2 b^2 c^2$.

The G. C. M. of the numerical coefficients is 6. The lowest power of a in any of the given expressions is a^2; the lowest power of b is b; the lowest power of c is c^2; and d is not a common factor. Therefore the required H. C. F. is $6\, a^2 bc^2$.

19. In general, to obtain the H. C. F. of two or more monomials:

Multiply the G. C. M. of their numerical coefficients by the product of their common literal factors, each to the lowest power to which it occurs in any of the given monomials.

20. Multinomial Expressions. — The method of finding the H. C. F. of multinomials by factoring is similar to that of finding the H. C. F. of monomials.

Ex. 1. The expressions

$$x^2 - 1 = (x-1)(x+1),$$

and $$x^2 + x - 2 = (x-1)(x+2),$$

have only the common factor $x - 1$. This is their H. C. F.

In general, *the H. C. F. of two or more multinomial expressions is the product of their common factors, each to the lowest power to which it occurs in any of them.*

EXERCISES VIII.

Find the H. C. F. of each of the following expressions:

1. ax^2, a^2x.
2. x^4, $12 x^2$.
3. $25 a^2b^2$, $10 ab^3$.
4. $6 x^2$, $4 x^3$, $10 x$.
5. $12 ab^2$, $6 a^2b$, $8 a^2b^3$.
6. $15 ax^2y^3$, $10 a^2x^3y^3$, $20 bx^2y^4$.
7. $20 a^2b^3x^4$, $30 a^3bx^3$, $24 ab^4x^2$.
8. $(a-b)^2$, $a^2 - b^2$.
9. $2 a^2 + 2 ab$, $3 ab + 3 b^2$.
10. $x^2 - 4 y^2$, $x^3 - 8 y^3$.
11. $x^3 - 9 xy^2$, $x^3 + 27 y^3$.
12. $(x^2y - xy^2)^2$, $xy(x^3 - y^3)$.
13. $2 x^3 - 2$, $2 x^3 + 4 x + 2$.
14. $ax^3 - ay^3$, $x^2 + xy + y^2$.
15. $x^2 - 9 y^2$, $x^2 + xy - 12 y^2$.
16. $2 x^3 - 50 x$, $x^2 - 3 x - 40$.
17. $x^3 + a^3$, $x^2 - ax - 2 a^2$.
18. $x^2 - ab + bx - ax$, $2 x^3 - 2 a^3$.
19. $x^2 + 3 x + 2$, $x^2 + 4 x + 3$.
20. $x^2 - 8 x + 15$, $x^2 - 9 x + 18$.
21. $x^2 - 10 x + 16$, $x^2 + 4 x - 12$.
22. $x^2 + xy - 30 y^2$, $x^2 - 2 xy - 15 y^2$.
23. $x^3 - x^2 - 42 x$, $3 x^2 + 9 x - 54$.

24. $x^3 - 3x^2 - 2x + 6,\ x^2 + 9x - 36.$
25. $x^2 + ab + ax + bx,\ x^2 - ab + ax - bx.$
26. $a^2 - (b-c)^2,\ (a-c)^2 - b^2.$
27. $a^2 - 2a,\ a^2 - 4,\ a^2 + 4a - 12.$
28. $x^3 - 27,\ x^2 - 6x + 9,\ x^2 + x - 12.$
29. $3x^2 - 3,\ 8x^3 - 8,\ x^2 + x - 2.$
30. $x^2 + 2x - 15,\ x^2 - 5x + 6,\ x^2 + x - 12.$
31. $x^2 - 3x - 40,\ x^2 + 3x - 10,\ x^2 - x - 30.$
32. $(b-c)^2 - a^2,\ (a+b)^2 - c^2,\ b^2 - (c-a)^2$

LOWEST COMMON MULTIPLES.

21. The **Lowest Common Multiple (L. C. M.)** of two or more integral algebraic expressions is the integral expression of lowest degree which is exactly divisible by each of them.

E.g., the L. C. M. of ax^2, bx^3, and cx^4 is evidently $abcx^4$

22. Ex. 1. Find the L. C. M. of a^3b, a^2bc^2, and ab^2c^4.

In the expression of lowest degree which is exactly divisible by each of the given expressions, the lowest power of a is evidently a^3, of b is b^2, and of c is c^4. Therefore their L. C. M. is $a^3b^2c^4$.

Observe that the power of each letter in the L. C. M. is the *highest* power to which it occurs in any of the given expressions. If the expressions contain numerical factors, the L. C. M. of these factors should be found as in Arithmetic.

Ex. 2. Find the L. C. M. of

$$3\,ab^2,\ 6\,b(x+y)^2,\ \text{and}\ 4\,a^2b(x-y)(x+y).$$

The L. C. M. of the numerical coefficients is 12.
The highest power of a in any of the expressions is a^2; of b is b^2; of $x+y$ is $(x+y)^2$; and of $x-y$ is $x-y$.
Consequently the required L. C. M. is $12\,a^2b^2(x+y)^2(x-y)$.

23. In general, to obtain the L. C. M. of two or more expressions:

Multiply the L. C. M. of their numerical coefficients by the product of all the different prime factors of the expressions, each to the highest power to which it occurs in any of them.

EXERCISES IX.

Find the L. C. M. of each of the following expressions:

1. $2a$, $3b$.
2. $4a^2b$, $2ab^2$, $3ax$.
3. $4ab$, $2a^2b^2$, $6a^3b$.
4. $7a^2b^2$, $14a^3bx$, $6a^2x^2$.
5. $7a^3m^2$, $21x^2m^3$, $2xm$.
6. $12a^3b^2x$, $18a^2bx^3$, ab^2x.
7. $2x^2$, x^2+x.
8. $3ab$, a^2-ab.
9. x^2-4, x^2+2x.
10. $(x+y)^2$, x^2-y^2.
11. $x^2(a-b)$, $x(a^3-b^3)$.
12. $4a-4b$, $2a^2x-2b^2x$.
13. x^2-16, x^2+x-20.
14. x^2-y^2, $x^2+3xy+2y^2$.
15. $4x^2-4$, x^3+5x^2-6x.
16. x^3+y^3, $(x^2+y^2)^2-x^2y^2$.
17. x^2+x-6, $x^2-8x+12$.
18. $x^2-12x+35$, $x^2+2x-35$.
19. x^2+6x+9, x^2+x-6.
20. $x^2-4x-12$, $x^2+12x+20$.
21. a, $2a+2b$, $(a+b)^2$.
22. x, $ax-bx$, a^2-b^2.
23. x^2+x-2, x^2+4x-5, $x^2+7x+10$.
24. $x^2-8x+15$, $x^2+3x-18$, $x^2-12x+27$.
25. $x^2-3x-10$, $x^2-4x-12$, $x^2-5x-14$.
26. $x^2-3x-40$, x^2-25, $x^2+10x+25$.
27. $2x^3-2y^3$, x^4-xy^3, $4x^3y+4y^4$.
28. $a(x+y)$, $a(x-y)$, $ab(x^2-y^2)$.
29. $3x+1$, $9x^2-1$, $27x^3+1$.
30. $x^2-(y+z)^2$, $y^2-(z+x)^2$, $z^2-(x+y)^2$

SOLUTION OF EQUATIONS BY FACTORING.

24. The roots of the equation
$$(x-1)(x-2) = 0 \qquad (1)$$
are evidently 1 and 2. For 1 reduces the first member to $0 \times (-1), = 0$; and 2 reduces the first member to $1 \times 0, = 0$. Therefore equation (1) is equivalent to the equations

$$x - 1 = 0 \text{ and } x - 2 = 0, \text{ jointly.}$$

This example illustrates the following method of solving an equation by factoring:

Transfer all terms to the first member. Factor this first member, and equate each of the resulting factors to zero. Solve the equations thus obtained.

Ex. 1. Solve the equation $x(x-2)(x+5) = 0$.

Equating factors to 0, $x = 0$; $x - 2 = 0$, whence $x = 2$; and $x + 5 = 0$, whence $x = -5$.

The roots are therefore 0, 2, and -5.

Ex. 2. Solve the equation $x^2 - 1 = 3$.

Transferring 3 to first member, and factoring, we have
$$(x-2)(x+2) = 0.$$

Equating factors to 0, $x - 2 = 0$, whence $x = 2$; and $x + 2 = 0$, whence $x = -2$.

The roots are therefore $+2$ and -2.

The statement $+2$ *and* -2 is usually written ± 2, read *positive and negative two.*

Ex. 3. Solve the equation $x^2 + 2x - 12 = 3$.

Transferring 3, $\qquad x^2 + 2x - 15 = 0$.

Factoring, $\qquad (x+5)(x-3) = 0$.

Equating factors to 0, $x + 5 = 0$, whence $x = -5$; and $x - 3 = 0$, whence $x = 3$.

The required roots are therefore $-5, 3$.

EXERCISES X.

Solve each of the following equations:

1. $x(x-1) = 0$.
2. $2x(x+2) = 0$.
3. $5(x-2)(x+3) = 0$.
4. $3(5x-6)(2x+3) = 0$.
5. $6x(x+3)(x-5) = 0$.
6. $5x(x^2-25) = 0$.
7. $x^2 - 7 = 2$.
8. $x^2 + 6 = 10$.
9. $x^2 + 5 = 30$.
10. $x^2 - 24 = 12$.
11. $2x^2 - 13 = 19$.
12. $3x^2 + 5 = 8$.
13. $5x^2 - 96 = 29$.
14. $23 - x^2 = 7$.
15. $3x^2 - 49 = 2x^2$.
16. $5x^2 + 11 = 4x^2 + 75$.
17. $x^2 - 3x + 2 = 0$.
18. $x^2 - x - 2 = 0$.
19. $x^2 + 5x + 6 = 0$.
20. $x^2 - 4x + 4 = 0$.
21. $x^2 - 5x + 6 = 0$.
22. $x^3 + x^2 - 2x = 0$.
23. $x^2 + 5x + 4 = 0$.
24. $x^2 + 6x + 9 = 0$.
25. $x^3 - 7x^2 + 12x = 0$.
26. $x^2 - 2x - 15 = 0$.
27. $x^2 + 10x + 24 = 0$.
28. $x^2 - 14x + 49 = 0$.

29. If 16 is added to the square of a number, the sum will be equal to eight times the number. What is the number?

30. If 24 is added to the square of a number, the sum will be equal to eleven times the number. What is the number?

31. In a number of 2 digits, the units' digit is 1 greater than the tens' digit. The product of the digits is equal to the number diminished by 25. What is the number?

32. The length of a field exceeds its breadth by 3 rods. If 18 rods were added to its length, and 2 rods were taken from its breadth, the area would be doubled. What are the dimensions of the field?

33. The number of square feet in the area of a square floor, increased by 12, is equal to seven times the number of feet in its side. What is the length of a side of the room?

CHAPTER VII.

FRACTIONS.

1. If, in division, the dividend be not a multiple of the divisor, the quotient is called a **Fraction**; as

$$a \div b;\ (ax^2 + 2bx) \div x^3.$$

2. The notation for a fraction in Algebra is the same as in ordinary Arithmetic.

Thus, $(ax^2 + 2bx) \div x^3$ is written $\dfrac{ax^2 + 2bx}{x^3}$.

3. As in Arithmetic, the dividend is called the **Numerator** of the fraction, the divisor the **Denominator**, and the two are called the **Terms** of the fraction.

4. An integer or an integral expression can be written in a *fractional form* with a denominator 1.

E.g., $\quad 7 = \dfrac{7}{1},\ a + b = \dfrac{a+b}{1}.$

5. By the definition of division, $\dfrac{a}{b}$ is a number which multiplied by b gives a; or

$$\frac{a}{b} \times b = a.$$

E.g., $\quad \dfrac{2}{3} \times 3 = 2.$

6. The Sign of a Fraction. — The sign of a fraction is written before the line separating its numerator from its denominator; as $+\dfrac{a}{b},\ -\dfrac{a}{b}.$

Reduction of Fractions to Lowest Terms.

7. A fraction is said to be *in its lowest terms* when its numerator and denominator have no common integral factor.

E.g., $\qquad \dfrac{2}{3}, \dfrac{x-1}{x^2+1}.$

8. Reduce $\frac{4}{6}$ to lowest terms. Let the value of $\frac{4}{6}$ be denoted by v; or $v = \frac{4}{6}$.

Multiplying by 6, $\quad 6v = \frac{4}{6} \times 6$, or $6v = 4$, by Art. 5.

Dividing by 2, $\qquad 3v = 2.$

Dividing by 3, $\qquad v = \frac{2}{3},$

But $\qquad\qquad\quad v = \frac{4}{6}.$

Therefore $\qquad \frac{4}{6} = \frac{2}{3} = \frac{4 \div 2}{6 \div 2}.$

This example illustrates the following principle:

The value of a fraction is not changed if both numerator and denominator be divided by the same number, not 0.

E.g., $\qquad \dfrac{a+ab}{a+ac} = \dfrac{(a+ab) \div a}{(a+ac) \div a} = \dfrac{1+b}{1+c}.$

9. Ex. 1. Reduce $\dfrac{6\,a^3b^2}{8\,a^2b^5}$ to its lowest terms.

The factor $2\,a^2b^2$ is the H. C. F. of the numerator and denominator. We therefore have

$$\dfrac{6\,a^3b^2}{8\,a^2b^5} = \dfrac{6\,a^3b^2 \div 2\,a^2b^2}{8\,a^2b^5 \div 2\,a^2b^2} = \dfrac{3\,a}{4\,b^3}.$$

A fraction is reduced to its lowest terms by dividing its numerator and denominator by the H. C. F. of its terms.

This step is called *cancelling common factors*, and can usually be done mentally, if the terms of the fraction are first resolved into their prime factors.

Ex. 2. $\qquad \dfrac{a^2 - x^2}{(a+x)^2} = \dfrac{(a+x)(a-x)}{(a+x)(a+x)} = \dfrac{a-x}{a+x}.$

FRACTIONS.

EXERCISES I.

Reduce each of the following fractions to its lowest terms:

1. $\dfrac{2}{6a}$.
2. $\dfrac{5a}{15a}$.
3. $\dfrac{4ab}{6a}$.
4. $\dfrac{14x^2}{21x}$.
5. $\dfrac{10a^3b}{15ab}$.
6. $\dfrac{11a^2b^2}{33a^2b^3}$.
7. $\dfrac{12x^3y^2}{18x^4y^2}$.
8. $\dfrac{25m^3n^4}{35m^5n^5}$.
9. $\dfrac{21a^3b^2c^4}{28a^4b^3c^2}$.
10. $\dfrac{22x^5y^4z}{33x^4y^5z^3}$.
11. $\dfrac{3}{3x-6}$.
12. $\dfrac{3a-6b}{6a-12b}$.
13. $\dfrac{x^2-x}{xy-y}$.
14. $\dfrac{a^2-ab}{a^2+ab}$.
15. $\dfrac{4(a^2-1)}{2ab+2b}$.
16. $\dfrac{a^2-1}{(a+1)^2}$.
17. $\dfrac{(m+2)^2}{m^2-4}$.
18. $\dfrac{(2x-y)^2}{4x^2-y^2}$.
19. $\dfrac{a^2-b^2}{a^2+2ab+b^2}$.
20. $\dfrac{9x^2-16y^2}{9x^2+24xy+16y^2}$.
21. $\dfrac{9m^2+30mn+25n^2}{9m^2-25n^2}$.
22. $\dfrac{6a^3+4a^2b}{12a^3-8a^2b}$.
23. $\dfrac{x^2-9}{x^2-4x+3}$.
24. $\dfrac{x^2-6x+5}{x^2-7x+6}$.
25. $\dfrac{a^2+6ab+9b^2}{a^2+4ab+3b^2}$.
26. $\dfrac{a+1}{a^3+1}$.
27. $\dfrac{a^3-1}{a^3+1}$.
28. $\dfrac{x^3-1}{x^4-x}$.
29. $\dfrac{x+2y}{x^3+8y^3}$.
30. $\dfrac{2a-3b}{8a^3-27b^3}$.

Reduction of Two or More Fractions to a Lowest Common Denominator.

10. Two or more fractions are said to have a common denominator when their denominators are the same.

E.g., $\dfrac{a}{b}$ and $\dfrac{c}{b}$; $\dfrac{x}{a^2-x^2}$ and $\dfrac{x-y}{(a+x)(a-x)}$.

The **Lowest Common Denominator** (L. C. D.) of two or more fractions is the L. C. M. of their denominators.

E.g., the L. C. D. of $\dfrac{a}{b^2c}$ and $\dfrac{d}{bc^2}$ is b^2c^2.

11. Change $\frac{3}{5}$ to an equivalent fraction whose denominator is 20. Let the value of the fraction $\frac{3}{5}$ be denoted by v, or $v=\frac{3}{5}$.

Multiplying by 5, $\quad\quad 5v = 3$.
Multiplying by 4, $\quad\ 20v = 12$.
Dividing by 20, $\quad\quad\ v = \frac{12}{20}$.
But $\quad\quad\quad\quad\quad\quad\ v = \frac{3}{5}$.
Therefore $\quad\quad\quad\ \frac{3}{5} = \frac{12}{20} = \frac{4\times 3}{4\times 5}$.

This example illustrates the following principle:

The value of a fraction is not changed if both numerator and denominator be multiplied by the same number, not 0.

E.g., $\quad \dfrac{a-x}{a+x} = \dfrac{(a-x)\times(a+x)}{(a+x)\times(a+x)} = \dfrac{a^2-x^2}{(a+x)^2}$.

12. Ex. 1. Reduce $\dfrac{a}{b^2c}$ and $\dfrac{d}{bc^2}$ to equivalent fractions having a lowest common denominator.

Their required L. C. D. is b^2c^2.

Multiplying both terms of $\dfrac{a}{b^2c}$ by $b^2c^2 \div b^2c, = c$, we have $\dfrac{ac}{b^2c^2}$; and both terms of $\dfrac{d}{bc^2}$ by $b^2c^2 \div bc^2, = b$, we have $\dfrac{bd}{b^2c^2}$.

Ex. 2. Reduce $x, = \dfrac{x}{1}$, and $\dfrac{y}{x-y}$ to equivalent fractions having a lowest common denominator.

The required L. C. D. is $x-y$.

Multiplying both terms of $\dfrac{x}{1}$ by $x-y$, we have $\dfrac{x^2-xy}{x-y}$; and both terms of $\dfrac{y}{x-y}$ by 1, we have $\dfrac{y}{x-y}$.

Ex. 3. Reduce $\dfrac{1}{x^2-3x+2}, = \dfrac{1}{(x-1)(x-2)}$, and $\dfrac{2}{x^2-1}, = \dfrac{2}{(x-1)(x+1)}$, to equivalent fractions having a lowest common denominator.

The required L. C. D. is $(x-1)(x-2)(x+1)$.

Multiplying both terms of the first fraction by

$$(x-1)(x-2)(x+1) \div (x-1)(x-2), = x+1,$$

we have
$$\frac{x+1}{(x-1)(x-2)(x+1)};$$

and both terms of the second fraction by

$$(x-1)(x-2)(x+1) \div (x-1)(x+1), = x-2,$$

we have
$$\frac{2x-4}{(x-1)(x-2)(x+1)}.$$

13. These examples illustrate the following method:

Take the L. C. M. of the denominators as the required denominator.

Divide this denominator by the denominator of each fraction; and multiply both numerator and denominator of the fraction by the quotient.

EXERCISES II.

1. Change $\dfrac{a}{3}$ to an equivalent fraction whose denominator is 12.

2. Change $\dfrac{a+b}{8}$ to an equivalemt fraction whose denominator is 24.

3. Change $\dfrac{ab}{c}$ to an equivalent fraction whose denominator is $2c^2$

4. Change $\dfrac{5ab - 7a^2b}{21\,a^2b^2c^4}$ to an equivalent fraction whose denominator is $63\,a^3b^3c^8$.

5. Change $\dfrac{a-b}{a+b}$ to an equivalent fraction whose denominator is a^2-b^2

6. Change $\dfrac{x+3}{x-2}$ to an equivalent fraction whose denominator is $2(x^2-4)$.

7. Change $\dfrac{x+y}{x-y}$ to an equivalent fraction whose denominator is $3(x^2-xy)$.

Reduce the following fractions to equivalent fractions having a lowest common denominator:

8. $1, \dfrac{a}{3}$.

9. $\dfrac{2a}{3}, \dfrac{3b}{6}$.

10. $\dfrac{5a^2b}{12}, \dfrac{7ab^2}{18}$.

11. $\dfrac{x-1}{15}, \dfrac{x+1}{25}$.

12. $\dfrac{2}{3a}, \dfrac{3}{5a}$.

13. $\dfrac{3}{4a^2}, \dfrac{5}{4a}$.

14. $\dfrac{2a}{3xy}, \dfrac{5b}{12x^2y}$.

15. $3, \dfrac{1}{a+1}$.

16. $\dfrac{5}{6}, \dfrac{3}{x-1}$.

17. $\dfrac{1}{x+1}, \dfrac{2}{3x+3}$.

18. $\dfrac{1}{x-y}, \dfrac{2}{(x-y)^2}$.

19. $\dfrac{5a}{2a-b}, \dfrac{7b}{2a^2-ab}$.

20. $\dfrac{3a}{a^2-b^2}, \dfrac{5}{a+b}$.

21. $\dfrac{1}{x^2-y^2}, \dfrac{5}{(x+y)^2}$.

22. $\dfrac{a+b}{(a-b)^2}, \dfrac{2a}{a^2-b^2}$.

23. $\dfrac{1}{x^2+2x+1}, \dfrac{7}{x^2-1}$.

24. $\dfrac{x-4}{x^2-3x+2}, \dfrac{2x+3}{x^2+x-2}$.

25. $\dfrac{x+2}{x^2-11x+30}, \dfrac{x-2}{x^2+x-30}$.

26. $\dfrac{1}{x+1}, \dfrac{2}{x-1}, \dfrac{3}{x^2-1}$.

27. $\dfrac{x+1}{2x-4}, \dfrac{x-1}{4x+8}, \dfrac{2x}{6x^2-24}$.

28. $\dfrac{x+1}{x^3-2x+1}, \dfrac{2x+1}{x^2-4x+3}, \dfrac{2x-1}{x^2-9}$.

Equations and Problems.

14. Ex. 1. Solve the equation $2x + \dfrac{x}{4} = 9$.

Multiplying by 4, $\quad 4 \times 2x + 4 \times \dfrac{x}{4} = 4 \times 9;$ (1)

or, since $4 \times \dfrac{x}{4} = x,$ $\qquad 8x + x = 36.$

Uniting terms, $\qquad 9x = 36.$

Dividing by 9, $\qquad x = 4.$

The step represented by (1) is called *clearing the equation of fractions*, and should be performed mentally.

To clear of fractions, we multiplied by 4, the denominator of the fractional term. If the equation contains more than one fraction, we multiply by their L. C. D.

Ex. 2. Solve the equation $\dfrac{x}{5} - \dfrac{2x-1}{3} = 3 - x$.

The L. C. D. is 15.

Multiplying by 15, $\quad 3x - 5(2x-1) = 15(3-x) \quad$ (1)

Removing parentheses, $\quad 3x - 10x + 5 = 45 - 15x. \quad$ (2)

Transferring terms, $\quad 3x - 10x + 15x = 45 - 5$.

Uniting terms, $\quad\quad\quad\quad\quad 8x = 40$.

Dividing by 8, $\quad\quad\quad\quad\quad x = 5$.

15. Observe that the sign of a fraction affects each term of the numerator, or the dividing line between the numerator and the denominator has the same effect as parentheses.

E.g.,
$$-\dfrac{a-b+c}{d} = -(a-b+c) \div d$$
$$= (-a+b-c) \div d$$
$$= \dfrac{-a+b-c}{d}.$$

Thus, in Ex. 2 Art. 14, the sign $-$ before the fraction $\dfrac{2x-1}{3}$ changes the signs of *both* terms in its numerator, and not simply the sign of the first term, when the denominator is removed. This caution should be kept in mind, and step (1) omitted in clearing of fractions.

16. Ex. Solve the equation $\dfrac{x+1}{6} - \dfrac{x-1}{8} = \dfrac{x+2}{12}$.

The L. C. D. is 24.

Multiplying by 24, $\quad 4x + 4 - 3x + 3 = 2x + 4$.

Transferring terms, $\quad 4x - 3x - 2x = 4 - 4 - 3$.

Uniting terms, $\quad\quad\quad\quad\quad x = -3$.

Dividing by -1, $\quad\quad\quad\quad\quad x = 3$.

17. Pr. In a number of two digits, the units' digit is two-thirds of the tens' digit. If the digits be interchanged, the resulting number will be 18 less than the given number. What is the number?

Let x stand for the tens' digit;

then $\frac{2}{3}x$ stands for the units' digit.

The given number is $10x + \frac{2}{3}x$,

and the resulting number is $10 \times \frac{2}{3}x + x, = \frac{20}{3}x + x$.

The problem states,

in *verbal* language: *the given number minus the resulting number is* 18;

in *algebraic* language: $10x + \frac{2}{3}x - (\frac{20}{3}x + x) = 18.$

Removing parentheses, $10x + \frac{2}{3}x - \frac{20}{3}x - x = 18.$

Clearing of fractions, $30x + 2x - 20x - 3x = 54.$

Uniting terms, $9x = 54.$

Dividing by 9, $x = 6,$

the tens' digit.

Then the units' digit is $\frac{2}{3}x, = 4.$

Therefore the required number is 64.

EXERCISES III.

Solve each of the equations:

1. $x + \dfrac{x}{2} = 6.$ 2. $x + \dfrac{x}{3} = 8.$ 3. $x - \dfrac{x}{4} = 6.$

4. $2x + \dfrac{x}{5} = 22.$ 5. $3x - \dfrac{x}{7} = 20.$ 6. $5x - \dfrac{x}{6} = 8.$

7. $2x + \dfrac{1}{2}x = 10.$ 8. $3x - \dfrac{2x}{3} = 14.$ 9. $\dfrac{x}{2} + \dfrac{x}{3} = 5.$

10. $\dfrac{x}{3} - \dfrac{x}{4} = 1.$ 11. $\dfrac{2x}{3} + \dfrac{3x}{4} = 34.$ 12. $\dfrac{7x}{8} - \dfrac{5x}{6} = 1.$

FRACTIONS.

13. $4x - \dfrac{2x}{3} + \dfrac{5x}{6} = 25.$
14. $\dfrac{2x}{3} + \dfrac{5x}{6} - \dfrac{x}{8} = 33.$
15. $5x - \dfrac{5x}{9} - \dfrac{3x}{12} = 151.$
16. $\dfrac{7x}{12} + \dfrac{3x}{8} - \dfrac{5x}{16} = 31.$
17. $x + \dfrac{2x}{3} - \dfrac{5x}{6} + \dfrac{x}{4} = 13.$
18. $3x - \dfrac{x}{5} + \dfrac{3x}{10} - \dfrac{2x}{15} = 89.$
19. $\dfrac{x+1}{3} + \dfrac{x}{2} = 7.$
20. $\dfrac{x-1}{3} + \dfrac{x}{4} = 9.$
21. $\dfrac{4x}{15} - \dfrac{x+5}{20} = 3.$
22. $\dfrac{3x}{5} - \dfrac{x-4}{6} = 5.$
23. $\dfrac{3x}{4} - \dfrac{x+6}{6} = 6.$
24. $\dfrac{5x}{8} - \dfrac{x+4}{12} = 4.$
25. $\dfrac{x-1}{4} + \dfrac{x+1}{6} = 17.$
26. $\dfrac{12-3x}{4} + \dfrac{11-2x}{3} = 1.$
27. $\dfrac{x-3}{8} - \dfrac{2x-5}{12} = -\dfrac{1}{3}.$
28. $\dfrac{x+1}{2} + \dfrac{x+2}{3} + \dfrac{x+3}{4} = 16.$
29. $\dfrac{x+1}{2} - \dfrac{x-2}{3} + \dfrac{x+3}{4} = 4.$
30. $\dfrac{3x+4}{5} - \dfrac{7x-3}{2} = \dfrac{x-16}{4}.$
31. $\dfrac{x-1}{10} + \dfrac{11x-1}{12} = \dfrac{5x-11}{4}.$
32. $\dfrac{x+1}{8} - \dfrac{x-4}{3} = \dfrac{x-3}{4} - \dfrac{x-5}{2}.$
33. $\dfrac{x-1}{2} + \dfrac{x-2}{3} = \dfrac{5x-1}{6} - \dfrac{x-3}{4}.$

Solve each of the following problems:

34. If ¼ of a number is diminished by 5, the remainder will be 2. What is the number?

35. Find the number whose double exceeds its half by 6.

36. A man paid $5600 for a house and lot. Five times the cost of the house is twice the cost of the lot. How much did he pay for each?

37. What time is it, if the number of hours since noon is equal to ½ of the number of hours till midnight?

38. Divide 60 into two parts, so that ½ of the greater shall be equal to ¾ of the less.

39. A son is ¼ as old as his father, and in 5 years he will be ⅓ as old. How old is each?

40. A son is ½ as old as his father, and 18 years ago he was ⅓ as old. How old is each?

41. A boy lost ¼ of his money, and afterwards found 8 cents. He then had ⅔ as much as at first. How much money had he at first?

42. Two men invest equal amounts. The first one loses $300, and the second one gains $300. The first then has only ⅔ as much as the second. How much did each invest?

43. Divide 80 into 3 parts, so that the second shall be 12 greater than the first, and the third ⅔ of the first and second.

44. Fifteen coins, dollars and quarter-dollars, amount to $7.50. How many coins of each kind are there?

45. A box contains a number of pencils, of which ⅓ are red, ¼ are blue, and 15 are black. How many pencils are red and how many are blue?

46. The deposits in a bank during 3 days amounted to $16,900. If the deposits each day after the first were ⅓ of those of the preceding day, how many dollars were deposited each day?

Addition and Subtraction of Fractions.

18. Add $\frac{2}{5}$ to $\frac{4}{5}$. We have

$$\tfrac{2}{5} + \tfrac{4}{5} = 2 \div 5 + 4 \div 5 = (2+4) \div 5 = \tfrac{2+4}{5}.$$

This example illustrates the following method of adding two or more fractions which have a common denominator:

The numerator of the sum is the sum of the numerators, and the denominator is the common denominator

A similar method is applied in subtracting fractions.

E.g., $\quad \dfrac{2x}{x-1} - \dfrac{1+x}{x-1} = \dfrac{2x-(1+x)}{x-1} = \dfrac{x-1}{x-1} = 1.$

19. If the fractions to be added or subtracted do not have a common denominator, they should first be reduced to equivalent fractions having a lowest common denominator.

Ex. 1. Simplify $\dfrac{a}{b^2c} + \dfrac{d}{bc^2}$.

We have $\dfrac{a}{b^2c} + \dfrac{d}{bc^2} = \dfrac{ac}{b^2c^2} + \dfrac{bd}{b^2c^2} = \dfrac{ac+bd}{b^2c^2}$.

Ex. 2. Simplify $\dfrac{2x-5y}{5} - \dfrac{3x-6y+2z}{4}$.

Reducing to L. C. D., we have

$$\dfrac{8x-20y}{20} - \dfrac{15x-30y+10z}{20}$$

$$= \dfrac{8x-20y-(15x-30y+10z)}{20}$$

$$= \dfrac{8x-20y-15x+30y-10z}{20} = \dfrac{-7x+10y-10z}{20}.$$

The expressions in this example are not *algebraic* fractions.

The beginner should be careful in subtracting a fraction to *change the sign of each term of the numerator*, and not that of the first term only.

In like manner we may change the sign of each term of the numerator (or denominator), if we change the sign of the fraction. Thus, in the result of Ex. 2, we have

$$\dfrac{-7x+10y-10z}{20} = -\dfrac{7x-10y+10z}{20}.$$

Ex. 3. Simplify $\dfrac{1}{x-1} - \dfrac{2}{x+1} + \dfrac{3x}{x^2-1}$.

The L. C. D. is x^2-1.

Therefore, $\dfrac{1}{x-1} - \dfrac{2}{x+1} + \dfrac{3x}{x^2-1} = \dfrac{x+1}{x^2-1} - \dfrac{2x-2}{x^2-1} + \dfrac{3x}{x^2-1}$

$$= \dfrac{x+1-2x+2+3x}{x^2-1}$$

$$= \dfrac{2x+3}{x^2-1}.$$

EXERCISES IV.

Simplify each of the following expressions:

1. $\dfrac{2}{a}+\dfrac{1}{a}.$
2. $\dfrac{a}{b}+\dfrac{c}{b}.$
3. $\dfrac{x+1}{x-3}+\dfrac{x-1}{x-3}.$
4. $\dfrac{1}{4a}+\dfrac{1}{12a}.$
5. $\dfrac{1}{ab}+\dfrac{1}{ac}.$
6. $\dfrac{1}{10a^2b}+\dfrac{1}{15ab^2}.$
7. $\dfrac{1}{1+x}+\dfrac{1}{1-x}.$
8. $\dfrac{2}{a-3}+\dfrac{3}{a+2}.$
9. $\dfrac{a+1}{a-1}+\dfrac{a-1}{a+1}.$
10. $\dfrac{2x+1}{3x-1}+\dfrac{1-2x}{3x+1}.$
11. $\dfrac{5}{a-2}+\dfrac{3}{a^2-4}.$
12. $\dfrac{2}{(2a-3)^2}+\dfrac{3}{4a^2-9}.$
13. $\dfrac{1}{x+1}+\dfrac{x-2}{x^2+3x+2}.$
14. $\dfrac{x-2}{x^2-25}-\dfrac{x-1}{x^2+3x-40}.$
15. $\dfrac{3}{x}-\dfrac{1}{x}.$
16. $\dfrac{5ab}{3x}-\dfrac{2ab}{3x}.$
17. $\dfrac{a+b}{2}-\dfrac{a-b}{2}.$
18. $\dfrac{a}{a-b}-\dfrac{b}{a-b}.$
19. $\dfrac{1}{4x}-\dfrac{1}{8x}.$
20. $\dfrac{2}{15a}-\dfrac{1}{10a^2}.$
21. $\dfrac{a}{a+b}-\dfrac{a}{a-b}.$
22. $\dfrac{2x+1}{2x-1}-\dfrac{2x-1}{2x+1}.$
23. $\dfrac{1}{1-x^2}-\dfrac{1}{1-x}.$
24. $\dfrac{1}{x^2-y^2}-\dfrac{1}{(x-y)^2}.$
25. $\dfrac{1}{(x+1)^2}-\dfrac{1}{(x-1)^2}.$
26. $\dfrac{1}{a}+\dfrac{1}{b}+\dfrac{1}{c}.$
27. $\dfrac{1}{xy}+\dfrac{1}{xz}+\dfrac{1}{yz}.$
28. $\dfrac{a}{bc}+\dfrac{b}{ac}+\dfrac{c}{ab}.$
29. $\dfrac{x-y}{xy}+\dfrac{y-z}{yz}-\dfrac{z-x}{xz}.$
30. $\dfrac{2x}{5}-\dfrac{3x}{10}+\dfrac{x}{4}.$
31. $\dfrac{3xy}{4}-\dfrac{5xy}{6}+\dfrac{xy}{8}.$
32. $\dfrac{a-3}{2}-\dfrac{a-5}{6}-\dfrac{4-a}{8}.$
33. $\dfrac{x-1}{2}-\dfrac{x-2}{3}+\dfrac{x+7}{6}.$
34. $\dfrac{3-2a}{3}+\dfrac{3a-2}{5}-\dfrac{6a+2}{10}.$
35. $\dfrac{5-3x}{4}-\dfrac{5x-4}{10}-\dfrac{25-19x}{15}.$

36. $\dfrac{1}{x-1} - \dfrac{1}{x+2} - \dfrac{3}{(x+2)^2}$.

37. $\dfrac{a}{x+a} - \dfrac{x}{x-a} + \dfrac{a^2}{x^2-a^2}$.

38. $\dfrac{1}{a-1} - \dfrac{1}{a+1} - \dfrac{2}{a^2-1}$.

39. $\dfrac{3+2x}{2-x} - \dfrac{2-3x}{2+x} - \dfrac{16x-x^2}{4-x^2}$.

40. $\dfrac{a+b}{a-b} - \dfrac{a-b}{a+b} + \dfrac{2(a^2+b^2)}{a^2-b^2}$.

41. $\dfrac{1}{x^2-y^2} + \dfrac{1}{(x+y)^2} + \dfrac{1}{(x-y)^2}$.

42. $\dfrac{3}{x^2+5x+6} - \dfrac{2}{x^2+6x+8}$.

43. $\dfrac{1}{x^2+9x+20} - \dfrac{1}{x^2+11x+30}$.

Reduction of Mixed Expressions to Improper Fractions.

20. A **Proper Fraction** is one whose numerator is of lower degree than its denominator in a common letter of arrangement.

E.g., $\dfrac{1}{x+1}$, $\dfrac{x-2}{x^2+2x-1}$.

An **Improper Fraction** is one whose numerator is of the same or of a higher degree than its denominator in a common letter of arrangement.

E.g., $\dfrac{x}{x+1}$, $\dfrac{x^3+3x^2+x-1}{x^2+2x-1}$.

If both integral and fractional terms occur in an expression, it is sometimes called a **Mixed Expression**.

21. Ex. 1. Reduce $a + \dfrac{b}{c}$ to an improper fraction. First reducing a to the form of a fraction with denominator c, we have

$$a + \dfrac{b}{c} = \dfrac{ac}{c} + \dfrac{b}{c} = \dfrac{ac+b}{c}.$$

This example illustrates the following method:

Multiply the integral part by the denominator of the fractional part.

To this product add algebraically the numerator of the fractional part, and write the sum as the required numerator.

110   ALGEBRA.   [Ch. VII

EXERCISES V.

Reduce each of the following mixed expressions to a fraction:

1. $a + \dfrac{b}{3}$.

2. $2x - \dfrac{3y}{4}$.

3. $x + \dfrac{x+4}{3}$.

4. $3a - \dfrac{4a-11}{5}$.

5. $2 - \dfrac{2x}{1+x}$.

6. $x - \dfrac{2xy - y^2}{x}$.

7. $x + 1 + \dfrac{1 - 2xy}{2xy}$.

8. $x + 5 - \dfrac{25 + 10x}{x+5}$.

9. $x - y + \dfrac{x^2 + y^2}{x + y}$.

10. $3x - 4 - \dfrac{2x^2 - 8}{x+2}$.

11. $2a - 1 + \dfrac{1}{2a+1}$.

12. $x - y + \dfrac{2xy}{x-y}$.

Reduction of Improper Fractions to Mixed Expressions.

22. Ex. 1. Reduce $\dfrac{2x^2 + x + 5}{x+1}$ to a mixed expression.

We have

$$\begin{array}{r|l} 2x^2 + x + 5 & x+1 \\ 2x^2 + 2x & 2x - 1 \\ \hline -x + 5 & \\ -x - 1 & \\ \hline 6 & \end{array}$$

But by Ch. III., Art. 40, we have

$$(2x^2 + x + 5) \div (x+1) = 2x - 1 + 6 \div (x+1),$$

or $\dfrac{2x^2 + x + 5}{x+1} = 2x - 1 + \dfrac{6}{x+1}$.

This example illustrates the following method:

Divide the numerator by the denominator, until the remainder is of lower degree than the divisor.

Write the remainder as the numerator of a fraction whose denominator is the divisor.

Add this fraction to the integral part of the quotient.

Ex. 2. Reduce $\dfrac{x^3 + x^2 - 4x + 3}{x^2 + 2x - 1}$ to a mixed expression.

We have

$$\begin{array}{r|l} x^3 + x^2 - 4x + 3 & x^2 + 2x - 1 \\ x^3 + 2x^2 - x & x - 1 \\ \hline -x^2 - 3x + 3 & \\ -x^2 - 2x + 1 & \\ \hline -x + 2 & \end{array}$$

Therefore, $\dfrac{x^3 + x^2 - 4x + 3}{x^2 + 2x - 1} = x - 1 + \dfrac{-x + 2}{x^2 + 2x - 1}$

$$= x - 1 - \dfrac{x - 2}{x^2 + 2x - 1}.$$

EXERCISES VI.

Reduce each of the following fractions to a mixed expression:

1. $\dfrac{3a - 2}{2}$.   2. $\dfrac{10x + 1}{5}$.   3. $\dfrac{a + b}{a}$.

4. $\dfrac{x^2 - y}{x}$.   5. $\dfrac{ax + b}{a}$.   6. $\dfrac{a^2}{a - 1}$.

7. $\dfrac{x^2 + 1}{x + 1}$.   8. $\dfrac{x^3 + 1}{x - 1}$.   9. $\dfrac{x^3 - 1}{x + 1}$.

10. $\dfrac{a^2x^2 - 2abx + b^2}{ax}$.   11. $\dfrac{3x^2 + 5x - 7}{x + 2}$.

12. $\dfrac{x^2 - 3x + 4}{x - 1}$.   13. $\dfrac{4x^3 + 6x^2 - 3}{x^2 + x - 1}$.

Multiplication of Fractions.

23. Multiply $\tfrac{2}{3}$ by $\tfrac{5}{7}$. Let the value of $\tfrac{2}{3}$ be denoted by v, and that of $\tfrac{5}{7}$ by w; or

$$v = \tfrac{2}{3}, \text{ and } w = \tfrac{5}{7}.$$

Multiplying the first equation by 3, and the second by 7, we have

$$3v = 2, \text{ and } 7w = 5.$$

112   ALGEBRA.   [Ch. VII

If equal numbers be multiplied by equal numbers, the products will be equal. Therefore,

$$3v \times 7w = 2 \times 5, \text{ or } 21\,vw = 10.$$

Dividing by 21, $\qquad vw = \frac{10}{21}.$

But $\qquad\qquad\qquad vw = \frac{2}{3} \times \frac{5}{7}.$

Therefore, $\qquad\quad \frac{2}{3} \times \frac{5}{7} = \frac{10}{21} = \frac{2\times 5}{3\times 7}.$

This example illustrates the following method of multiplying fractions:

The numerator of the product is the product of the numerators; and the denominator of the product is the product of the denominators.

24. Ex. 1. Find the product $\dfrac{15\,a^3b^2}{22\,x^2y^5} \times \dfrac{14\,xy^2}{25\,a^2b}.$

The factor 5 is common to the numerator of the first fraction and the denominator of the second. Since to cancel a common factor *before* multiplication is equivalent to cancelling it *after* the multiplication, we should first cancel 5. For a similar reason we should cancel the factors 2, a^2, b, x, and y^2 before the multiplication. We then have

$$\frac{3\,ab}{11\,xy^3} \times \frac{7}{5} = \frac{21\,ab}{55\,xy^3}.$$

In general, if the numerator of one fraction and the denominator of another have common factors, such factors should be cancelled before the multiplication is performed.

Ex. 2. Find the product $\dfrac{8\,a^2}{a^2-b^2} \times \dfrac{(a+b)^2}{4\,a}.$

Cancelling the common factors, $4\,a$ and $a+b$, we have

$$\frac{2\,a}{a-b} \times \frac{a+b}{1} = \frac{2\,a(a+b)}{a-b}.$$

25. Ex. Find the product $\dfrac{x-y}{x^2+y^2} \times (x+y)$.

We have $\dfrac{x-y}{x^2+y^2} \times \dfrac{x+y}{1} = \dfrac{x^2-y^2}{x^2+y^2}$.

Observe that *a fraction is multiplied by an integer, by multiplying its numerator by the integer.*

26. If one of the factors is a mixed expression, it should first be reduced to an improper fraction.

Ex. Find the product $\left(1-\dfrac{1}{x}\right)\left(\dfrac{1}{x^2-1}\right)$.

We have $\left(1-\dfrac{1}{x}\right)\left(\dfrac{1}{x^2-1}\right) = \dfrac{x-1}{x} \times \dfrac{1}{x^2-1}$

$$= \dfrac{1}{x} \times \dfrac{1}{x+1} = \dfrac{1}{x(x+1)}.$$

EXERCISES VII.

Simplify the following expressions:

1. $\dfrac{3}{a} \times 2$.
2. $\dfrac{a}{5b} \times 5$.
3. $\dfrac{a}{12y} \times 8y$.

4. $\dfrac{3x}{5y^2} \times 15y^3$.
5. $\dfrac{4a}{3b} \times \dfrac{3b}{8c}$.
6. $\dfrac{4ax}{3} \times \dfrac{6b}{xy}$.

7. $\dfrac{3a^2}{4b^2} \times \dfrac{2b}{3a}$.
8. $\dfrac{25x^2}{16a} \times \dfrac{8a^2}{15x}$.

9. $\dfrac{14a^3b}{15xy^2} \times \dfrac{10x^2y}{21ab^3}$.
10. $\dfrac{2a}{3b} \times \dfrac{5b}{8c} \times \dfrac{3x}{4}$.

11. $\dfrac{a}{b^2c^2} \times \dfrac{bcd}{ax} \times \dfrac{bx}{ad^2}$.
12. $\dfrac{3a-3b}{5ab} \times \dfrac{25a^2b^2}{9}$.

13. $\dfrac{a-1}{3} \times (a+2)$.
14. $\dfrac{a+1}{a-1} \times (a-2)$.

15. $\dfrac{x-1}{10x} \times \dfrac{1+x^2}{x^2-1}$.
16. $\dfrac{x^2+3x+2}{6x^2} \times \dfrac{3xy}{x+1}$.

17. $\dfrac{a+1}{5a-3b} \times \dfrac{35a^2-21ab}{a^2+a^3}$.
18. $\dfrac{a^2}{a^2-16} \times \dfrac{a^2+8a+16}{a^3-a^2}$.

19. $\dfrac{x^2+xy}{(x-y)^2} \times \dfrac{x^2-y^2}{x^2+4xy+3y^2}$.
20. $\dfrac{x^2-1}{x^2} \times \dfrac{x}{x+1} \times \dfrac{3}{x-1}$.

21. $\dfrac{a^2+2a+1}{x^2+3x+2} \times \dfrac{x^2-4}{a^3-1} \times \dfrac{a+1}{x-2}$.

22. $\dfrac{5x^3-5x}{x^2-25} \times \dfrac{x+1}{x-2} \times \dfrac{x^2+3x-10}{x^3-1}$.

23. $\dfrac{x+2}{x^2-5x+4} \times \dfrac{x^2-16}{x-5} \times \dfrac{x^2-6x+5}{x^2+3x+2}$.

24. $\left(\dfrac{1}{b}-\dfrac{1}{a}\right)\dfrac{2ab}{a^2-b^2}$.

25. $\left(1-\dfrac{x^2}{4}\right)\dfrac{16x^2}{6-x-x^2}$.

Reciprocal Fractions.

27. The **Reciprocal** of a fraction is a fraction whose numerator is the denominator, and whose denominator is the numerator of the given fraction.

E.g., the reciprocal of $\dfrac{a}{b}$ is $\dfrac{b}{a}$.

28. The product of a fraction and its reciprocal is 1.

For $\qquad \dfrac{a}{b} \times \dfrac{b}{a} = \dfrac{ab}{ba} = 1$.

Division of Fractions.

29. Divide $\tfrac{2}{3}$ by $\tfrac{5}{7}$.

Let the value of $\dfrac{\tfrac{2}{3}}{\tfrac{5}{7}}$ be denoted by v; or, $v = \dfrac{\tfrac{2}{3}}{\tfrac{5}{7}}$.

Multiplying by $\tfrac{5}{7}$, $\qquad \tfrac{5}{7} v = \tfrac{2}{3}$.

Multiplying by $\tfrac{7}{5}$, $\qquad \tfrac{7}{5} \times \tfrac{5}{7} v = \tfrac{2}{3} \times \tfrac{7}{5}$;

or, since $\tfrac{7}{5} \times \tfrac{5}{7} = 1$, $\qquad v = \tfrac{2}{3} \times \tfrac{7}{5}$.

But $v = \dfrac{\tfrac{2}{3}}{\tfrac{5}{7}}$; therefore $\dfrac{\tfrac{2}{3}}{\tfrac{5}{7}} = \tfrac{2}{3} \times \tfrac{7}{5}$.

This example illustrates the following method of dividing one fraction by another:

Multiply the dividend by the reciprocal of the divisor.

30. Ex. $\dfrac{4(a^2-ab)}{(a+b)^2} \div \dfrac{6a}{a^2-b^2} = \dfrac{4a(a-b)}{(a+b)^2} \times \dfrac{(a-b)(a+b)}{6a}$

$\qquad\qquad\qquad = \dfrac{2(a-b)^2}{3(a+b)}$.

FRACTIONS.

31. Ex. $\dfrac{x^2-y^2}{x^2+y^2} \div (x-y) = \dfrac{x^2-y^2}{x^2+y^2} \div \dfrac{x-y}{1}$

$= \dfrac{x^2-y^2}{x^2+y^2} \times \dfrac{1}{x-y} = \dfrac{x+y}{x^2+y^2}.$

Observe that *a fraction is divided by an integer by multiplying its denominator by the integer.*

EXERCISES VIII.

Simplify the following expressions:

1. $\dfrac{8}{a} \div 4.$ 2. $\dfrac{a}{x} \div 2.$ 3. $a \div \dfrac{2}{3}.$

4. $7x \div \dfrac{x}{7}.$ 5. $ab \div \dfrac{a}{b}.$ 6. $\dfrac{a}{b} \div \dfrac{a}{x}.$

7. $\dfrac{a}{x} \div \dfrac{b}{x}.$ 8. $\dfrac{7}{a} \div \dfrac{5}{a}.$ 9. $\dfrac{3\,ab}{5\,xy} \div \dfrac{2\,ab}{7\,xy}.$

10. $\dfrac{6\,a^2}{5\,b^2} \div \dfrac{3\,a}{15\,b}.$ 11. $\dfrac{x^2y}{a} \div \dfrac{xy}{a^2}.$

12. $\dfrac{4\,a^2x}{21\,b^2y^2} \div \dfrac{6\,a^2x}{35\,b^2y}.$ 13. $\dfrac{ab+ac}{b} \div \dfrac{b+c}{b^2}.$

14. $\dfrac{3x+6}{5} \div \dfrac{x+2}{15}.$ 15. $\dfrac{5x^2-10x}{6a} \div \dfrac{x-1}{12\,a^2}.$

16. $\dfrac{x^2-9}{6x} \div \dfrac{x+3}{2}.$ 17. $\dfrac{a^3-a^2x}{a^3+a^2x} \div \dfrac{a^4-a^2x^2}{a^4+a^2x^2}.$

18. $\dfrac{x^2+5x-6}{2\,xy} \div \dfrac{x-3}{3\,xy}.$ 19. $\dfrac{a-b}{3\,a^2-9\,ab} \div \dfrac{a^3b^2-a^2b^3}{a^2-9\,b^2}.$

20. $\dfrac{x^2-3x+2}{x^2-5x+6} \div \dfrac{x^2-4x+4}{x^2-9}.$ 21. $\dfrac{a^3+1}{5\,a^2b^2} \div \dfrac{(a+1)^2}{10\,a^2b^2}.$

22. $\dfrac{x^3-y^3}{x+y} \div \dfrac{x^2+xy+y^2}{x^2-y^2}.$ 23. $\dfrac{8\,m^3-n^3}{(x-y)^2} \div \dfrac{4\,m^2-n^2}{x^2-y^2}.$

24. $\left(a+2-\dfrac{1}{a-1}\right) \div \dfrac{a^2-9}{a^2-1}.$ 25. $\left(x+1-\dfrac{x-1}{x+1}\right) \div \dfrac{x^2+x+2}{x^2+3x+2}.$

Complex Fractions.

32. A Complex Fraction is a fraction whose numerator and denominator, either or both, are fractions.

E.g., $\dfrac{\frac{2}{3}}{\frac{4}{5}}, \quad \dfrac{\frac{a+x}{a-x}}{\frac{a+y}{a-y}}, \quad \dfrac{1+\frac{1}{x}}{1-\frac{1}{x}}.$

Observe that the line which separates the terms of the complex fraction is drawn heavier than the lines which separate the terms of the fractions in its numerator and denominator.

33. Ex. 1. Simplify $\dfrac{\frac{1-x^2}{x}}{1-x}.$

Multiplying both numerator and denominator by x, we obtain

$$\dfrac{\frac{x(1-x^2)}{x}}{x(1-x)} = \dfrac{1-x^2}{x(1-x)} = \dfrac{1+x}{x}.$$

To reduce a complex fraction to a simple fraction:

Multiply both its terms by the L. C. D. of the fractions in the numerator and denominator.

Ex. 2. Simplify $\dfrac{a-\frac{ax}{a+x}}{a+\frac{ax}{a-x}}.$

The L. C. D. is $(a+x)(a-x)$. Multiplying both terms of the fraction by $(a+x)(a-x)$, we have

$$\dfrac{a(a+x)(a-x) - ax(a-x)}{a(a+x)(a-x) + ax(a+x)} = \dfrac{a(a^2-x^2) - ax(a-x)}{a(a^2-x^2) + ax(a+x)}$$

$$= \dfrac{a^3 - ax^2 - a^2x + ax^2}{a^3 - ax^2 + a^2x + ax^2}$$

$$= \dfrac{a^3 - a^2x}{a^3 + a^2x} = \dfrac{a-x}{a+x}.$$

EXERCISES IX.

Simplify each of the following complex fractions:

1. $\dfrac{\dfrac{a}{3}}{\dfrac{a}{3}-\dfrac{b}{3}}.$

2. $\dfrac{1+\dfrac{1}{x}}{1-\dfrac{1}{x}}.$

3. $\dfrac{a+\dfrac{b}{3}}{a-\dfrac{b}{3}}.$

4. $\dfrac{\dfrac{1}{a}+\dfrac{1}{b}}{\dfrac{1}{a}-\dfrac{1}{b}}.$

5. $\dfrac{1-\dfrac{1}{a+1}}{1+\dfrac{1}{a-1}}.$

6. $\dfrac{\dfrac{a-b}{a^2-b^2}}{\dfrac{a-b}{a+b}}.$

7. $\dfrac{a+1+\dfrac{1}{a-1}}{a-1+\dfrac{1}{a+1}}.$

8. $\dfrac{x-\dfrac{xy}{x+y}}{x+\dfrac{xy}{x-y}}.$

9. $\dfrac{x+y-\dfrac{2xy}{x+y}}{x-y+\dfrac{2xy}{x-y}}.$

10. $\dfrac{\dfrac{1}{a+1}+\dfrac{1}{a-1}}{\dfrac{1}{a-1}-\dfrac{1}{a+1}}.$

11. $\dfrac{x+2-\dfrac{12}{x+1}}{x-3-\dfrac{32}{x+1}}.$

12. $\dfrac{x+5+\dfrac{6}{x}}{x+1-\dfrac{2}{x}}.$

13. $\dfrac{x^2-1-\dfrac{6}{x^2}}{x^2+1-\dfrac{2}{x^2}}.$

14. $\dfrac{1-\dfrac{1}{x}-\dfrac{56}{x^2}}{x-\dfrac{2}{x}-\dfrac{48}{x^2}}.$

CHAPTER VIII.

FRACTIONAL EQUATIONS IN ONE UNKNOWN NUMBER.

1. A **Fractional Equation** is an equation whose members, either or both, are fractional expressions in the unknown number or numbers.

E.g., $\quad \dfrac{3}{x+2} = \dfrac{2}{x+1}, \quad x - 2 + \dfrac{4-2x}{x+1} = 0.$

2. Ex. 1. Solve the equation $\quad \dfrac{3}{x+2} = \dfrac{2}{x+1}.$

Multiplying by $(x+1)(x+2)$, $\quad 3(x+1) = 2(x+2)$. $\quad$ (1)

Transferring terms, $\quad 3x - 2x = 4 - 3.$

Uniting terms, $\quad x = 1.$

Check: $\quad \dfrac{3}{1+2} = \dfrac{2}{1+1},$ or $1 = 1.$

In clearing this equation of fractions, we multiplied by an expression, $(x+1)(x+2)$, which contains the unknown number. In such a case a root may be introduced. But it is proved in School Algebra, Ch. X., that, if a root is introduced in clearing of fractions, it must be a root of one of the factors of the L. C. D. equated to 0. Since 1 is not a root of

$$x + 1 = 0, \text{ or of } x + 2 = 0,$$

it is a root of the given equation.

Ex. 2. Solve the equation $\dfrac{2x+19}{5x^2-5} - \dfrac{17}{x^2-1} = -\dfrac{3}{x-1}.$

The L.C.D. is $5(x^2 - 1)$, $= 5(x-1)(x+1).$

118

FRACTIONAL EQUATIONS.

Multiplying by $5(x^2-1)$, $2x+19-85=-15x-15$.
Transferring terms, $\quad 2x+15x=-15-19+85$.
Uniting terms, $\quad 17x=51$.
Dividing by 17, $\quad x=3$.

Since 3 is not a root of $x-1=0$, or of $x+1=0$, it is a root of the given equation.

Ex. 3. Solve the equation $\dfrac{6x+1}{4}-\dfrac{2x-1}{3x-2}=\dfrac{3x-1}{2}$.

When the denominators of some of the fractions do not contain the unknown number, it is usually better first to unite these fractions.

Transferring $\dfrac{3x-1}{2}$, $\quad \dfrac{6x+1}{4}-\dfrac{3x-1}{2}-\dfrac{2x-1}{3x-2}=0$.

Uniting first two fractions, $\quad \dfrac{3}{4}-\dfrac{2x-1}{3x-2}=0$.

Multiplying by $4(3x-2)$, $\quad 9x-6-8x+4=0$.

Transferring and uniting terms, $\quad x=2$.

Since 2 is not a root of $3x-2=0$, it is a root of the given equation.

EXERCISES I.

Solve each of the following equations:

1. $\dfrac{x+1}{x-7}=9$. 2. $\dfrac{x-5}{x-9}=2$. 3. $\dfrac{2x-3}{x-3}=3$.

4. $\dfrac{5x-4}{x+3}=2$. 5. $\dfrac{x-3}{x-4}=\dfrac{x-5}{x-7}$. 6. $\dfrac{x-3}{x-5}=\dfrac{x-6}{x-9}$.

7. $\dfrac{x-4}{x-5}=\dfrac{x-8}{x-11}$. 8. $\dfrac{x-4}{2x-11}=\dfrac{x-2}{2x-5}$.

9. $\dfrac{5x-3}{2x+5}=\dfrac{5x-7}{2x+3}$. 10. $\dfrac{x+1}{4}-\dfrac{x-1}{6x-12}=\dfrac{x-1}{3}$.

11. $\dfrac{2x+3}{4}-\dfrac{x-1}{6x-8}=\dfrac{x+2}{2}$. 12. $\dfrac{2x+1}{5}+\dfrac{3x-2}{6x+3}=\dfrac{6x-1}{15}$.

13. $\dfrac{5x}{3x-2}-\dfrac{7x+2}{9x-6}=1$. 14. $\dfrac{3x-1}{2x-3}-\dfrac{x+9}{4x-6}=1$.

15. $\dfrac{3x-5}{x-1} - \dfrac{2x-5}{x-2} = 1.$ 16. $\dfrac{9x-7}{3x-2} - \dfrac{4x-5}{2x-3} = 1.$

17. $\dfrac{2x-7}{x-2} + \dfrac{3x-7}{x-3} = 5.$ 18. $\dfrac{2}{x-1} + \dfrac{3}{x-2} = \dfrac{20}{4x-7}.$

19. $\dfrac{9}{x-5} + \dfrac{28}{45-7x} = \dfrac{5}{x-9}.$ 20. $\dfrac{x+1}{x-1} - \dfrac{x-1}{x+1} = \dfrac{8}{x^2-1}.$

21. $\dfrac{x+2}{x-2} - \dfrac{x-2}{x+2} = \dfrac{16}{x^2-4}.$ 22. $\dfrac{1+2x}{1-2x} - \dfrac{1-2x}{1+2x} = \dfrac{10-2x}{1-4x^2}.$

23. $\dfrac{3x-5}{3x-3} - \dfrac{2x-7}{2x+2} = \dfrac{19x-3}{6x^2-6}.$ 24. $\dfrac{2x+3}{2x-3} - \dfrac{2x-3}{2x+3} = \dfrac{25x-3}{4x^2-9}.$

Problems.

3. Pr. 1. The value of a fraction when reduced to its lowest terms is $\tfrac{1}{5}$. If its numerator and denominator be each diminished by 1, the resulting fraction will be equal to $\tfrac{1}{6}$. What is the fraction?

The numerator of the required fraction must be a multiple of 1, and the denominator the same multiple of 5.

Let x stand for this multiple. The required fraction is $\dfrac{x}{5x}$.

The problem states,

in *verbal* language: $\quad \dfrac{\text{numerator minus 1}}{\text{denominator minus 1}} = \dfrac{1}{6}.$

in *algebraic* language: $\quad \dfrac{x-1}{5x-1} = \dfrac{1}{6}.$

Whence, $\qquad x = 5.$

The required fraction is $\qquad \dfrac{x}{5x} = \dfrac{5}{25}.$

Pr. 2. A number of men received $120, to be divided equally. If their number had been 4 less, each one would have received three times as much. How many men were there?

Let x stand for the number of men. Then each man received $\dfrac{120}{x}$ dollars. If their number had been 4 less, each one would have received $\dfrac{120}{x-4}$ dollars.

The problem states,

in *verbal* language: *The number of dollars each would have received, if there had been four less, is equal to three times the number of dollars each received;*

in *algebraic* language: $\dfrac{120}{x-4} = 3 \times \dfrac{120}{x}$.

Whence, $x = 6$.

Therefore there were six men.

Pr. 3. A can do a piece of work in 9 days, B in 6 days, and A, B, and C together in 3 days. In how many days can C do the work?

Let x stand for the number of days it takes C to do the work. Then, in one day,

A does $\dfrac{1}{9}$ of the work; B does $\dfrac{1}{6}$; and C does $\dfrac{1}{x}$

In 3 days,

A does $\dfrac{3}{9}$ of the work; B does $\dfrac{3}{6}$; and C does $\dfrac{3}{x}$.

Therefore, in 3 days, A, B, and C together do

$$\dfrac{3}{9} + \dfrac{3}{6} + \dfrac{3}{x} \text{ of the work.}$$

The problem states,

in *verbal* language: *The work done by A, B, and C together in 3 days is equal to all the work*, or 1;

in *algebraic* language: $\dfrac{3}{9} + \dfrac{3}{6} + \dfrac{3}{x} = 1$.

Whence, $x = 18$.

Therefore C can do the work in 18 days.

Pr. 4. A cistern has 3 taps. By the first it can be emptied in 80 minutes, by the second in 200 minutes, and by the third in 5 hours. After how many hours will the cistern be emptied, if all the taps are opened?

Let x stand for the number of minutes it takes the three taps together to empty the cistern.

Then, in 1 minute, the three together will empty $\frac{1}{x}$ of the cistern.

But, in 1 minute, the first will empty $\frac{1}{80}$ of the cistern; the second $\frac{1}{200}$, and the third $\frac{1}{300}$; and together they will empty $\frac{1}{80} + \frac{1}{200} + \frac{1}{300}$ of the cistern.

Therefore $$\frac{1}{80} + \frac{1}{200} + \frac{1}{300} = \frac{1}{x}.$$

Whence $x = 48.$

It will take the three taps together 48 minutes, or $\frac{4}{5}$ of an hour, to empty the cistern.

EXERCISES II.

1. If a certain number diminished by 1 is divided by the same number increased by 1, the quotient will be equal to $\frac{2}{3}$. What is the number?

2. A boy being asked his age replied: "If three times the number of years in my age is diminished by 16 and the result is divided by the number of years in my age, this quotient will be equal to $\frac{5}{3}$." How old was he?

3. Two equal fractions are formed. The numerators are formed by diminishing a certain number by 1 and 4 respectively. The corresponding denominators are formed by diminishing the same number by 3 and 2 respectively. What is the number?

4. The numerator of a fraction is formed by subtracting 1 from a certain number, and the denominator by adding 1 to the same number. If this fraction is added to the reciprocal of the number, the result will be $\frac{5}{6}$. What is the number?

5. The difference of two numbers is 3. Eight divided by the greater is equal to 5 divided by the less. What are the numbers?

6. If the reciprocal of a certain number is subtracted from the reciprocal of 5, the remainder will be equal to $\frac{2}{85}$. What is the number?

FRACTIONAL EQUATIONS.

7. A steamer can run 20 miles an hour in still water. If it can run 72 miles with the current in the same time that it can run 48 miles against the current, what is the speed of the current?

8. A can do a piece of work in 2 days, B in 3 days, and C in 6 days. In how many days can A, B, and C do the work?

9. A cistern has 3 pipes. The first can fill the cistern in 5 hours and the second in 5 hours; the third can empty it in 15 hours. How long would it take to fill the cistern if the 3 pipes were open?

10. A can do a piece of work in $2\frac{1}{2}$ days, B in $1\frac{1}{2}$ days, and C in $1\frac{2}{3}$ days. In what time can A, B, and C do the work?

11. A number has three digits, which increase by 1 from left to right. The quotient of the number divided by the sum of the digits is 26. What is the number?

12. A man being asked his age replied: "My age 3 years ago divided into 75 is equal to my present age divided into 84." What is his present age?

13. A man wished to give to each of 23 men and women the same amount. He divided $39 among the men and $30 among the women. How many of each were there?

14. A farmer intended to feed 80 bushels of corn to a certain number of sheep. When 6 of the sheep died, he could have sold 24 bushels of corn and have had enough left to give each remaining sheep the same amount as before. How many sheep had he?

15. There are three consecutive numbers. The sum of the quotient of 1 divided by the first and 2 divided by the third, diminished by the quotient of 3 divided by the second, is equal to 0. What are the numbers?

NOTE. — Assume the numbers to be $x-1$, x, $x+1$.

16. In one hour a train runs 10 miles further than a man rides on a bicycle in the same time. If it takes the train 6 hours longer to run 255 miles than it takes the man to ride 63 miles, what is the rate of the train?

CHAPTER IX.

LITERAL EQUATIONS IN ONE UNKNOWN NUMBER.

1. The unknown numbers of an equation are frequently to be determined in terms of general numbers, *i.e.*, in terms of numbers represented by letters. The latter are commonly represented by the leading letters of the alphabet, *a, b, c*, etc.

Such numbers as *a, b, c*, etc., are to be regarded as known.

E.g., in the equation $x + a = b$, a and b are the *known* numbers, and x is the *unknown* number.

From this equation we obtain $x = b - a$.

2. A **Numerical Equation** is one in which all the known numbers are numerals; as $2x + 3 = 7$; $4x - 3y = 7$.

A **Literal Equation** is one in which some or all of the known numbers are literal; as $2ax + 3b = 5$; $ax + by = c$.

3. Ex. 1. Solve the equation $\dfrac{x-a}{b} + \dfrac{x-b}{a} = -\dfrac{(a-b)^2}{2ab}$.

Clearing of fractions,

$$2ax - 2a^2 + 2bx - 2b^2 = -a^2 + 2ab - b^2$$

Transferring and uniting terms,

$$2(a+b)x = a^2 + 2ab + b^2.$$

Dividing by $2(a+b)$, $\quad x = \dfrac{a+b}{2}.$

Notice that the above equation, although algebraically **fractional**, is integral in the unknown number x. The equation which follows is fractional in the unknown number.

Ex. 2. Solve the equation $\dfrac{a+x}{b+x} = \dfrac{a+1}{b+1}.$

Multiplying by $(b+x)(b+1)$, $(a+x)(b+1) = (b+x)(a+1).$
Simplifying, $\qquad ab + bx + a + x = ab + ax + b + x.$
Cancelling terms, $\qquad bx + a = ax + b.$
Transferring and uniting terms, $(b - a)x = b - a.$
Dividing by $b - a,$ $\qquad x = 1.$

EXERCISES I.

Solve each of the following equations:

1. $x - a = b.$
2. $mx + a = b.$
3. $a - bx = c.$
4. $mx + nx = 2.$
5. $x - ax = a - 1.$
6. $m(x - n) = n(x - m).$
7. $a(x - n) = b(n - x).$
8. $a(x + a) - b(x - b) = 2ax + (a - b)^2$
9. $a + \dfrac{b}{x} = c.$
10. $\dfrac{a}{c} + b = \dfrac{b}{x} + a.$
11. $\dfrac{x+a}{x-a} = \dfrac{5}{4}.$
12. $\dfrac{a}{b} = \dfrac{x - b^2}{x - a^2}.$
13. $\dfrac{b^2}{ax} + \dfrac{b}{a} = \dfrac{a}{b} + \dfrac{a}{x}$
14. $\dfrac{x+a}{2} - \dfrac{2}{x+a} = \dfrac{x-a}{2}.$
15. $x - \dfrac{x-a}{2} = x + \dfrac{x-c}{3}.$
16. $m - \dfrac{x}{m} = a\left(\dfrac{n}{a} + \dfrac{x}{n}\right).$

General Problems.

4. A **General Problem** is one in which the *known* numbers are literal.

Pr. 1. The greater of two numbers is m times the less, and their sum is s. What are the numbers?

Let x stand for the less required number. Then mx stands for the greater. By the condition of the problem, we have

$$x + mx = s;$$

whence, $x = \dfrac{s}{1+m}$, the less number, and $mx = \dfrac{ms}{1+m}$, the greater.

If $m = 3$ and $s = 84$, we have

$$x = \frac{84}{1+3} = 21, \text{ and } mx = 3 \times 21 = 63.$$

When the numbers are equal, $m = 1$, and we obtain

$$x = \frac{s}{2}, \text{ and } mx = \frac{s}{2},$$

for all values of s; that is, either of the two numbers is half their sum.

Thus the solution of this general problem includes the solutions of all like problems. A solution for any like problem is obtained by substituting particular values for m and s, as above.

Pr. 2. A cistern has two taps. By the first it can be filled in a minutes, and by the second in b minutes. How many minutes will it take the two taps together to fill the cistern?

Let x stand for the number of minutes it takes the two taps to fill the cistern. Then, in 1 minute, the two together will fill $\frac{1}{x}$ of the cistern.

But, in 1 minute, the first will fill $\frac{1}{a}$ of the cistern, the second $\frac{1}{b}$; and together they will fill $\frac{1}{a} + \frac{1}{b}$ of the cistern.

Therefore $$\frac{1}{a} + \frac{1}{b} = \frac{1}{x}.$$

Whence $$x = \frac{ab}{a+b}.$$

This solution gives a general rule for solving problems of like character. In a particular example, a may be the number of minutes it takes a tap to fill a cistern, the number of hours it takes a man to build a wall, to dig a ditch, to plough a field, etc.

Pr. 3. If one man can dig a ditch in 6 days, and a second man in 3 days, in how many days can they dig the ditch, working together?

LITERAL EQUATIONS.

Substituting $a=6$, $b=3$, in the result of Pr. 2, we have

$$x = \tfrac{6 \times 3}{6+3} = 2.$$

Therefore together they can dig the ditch in 2 days.

EXERCISES II.

1. Find a number, such that the result of adding it to n shall be equal to n times the number. Let $n=2$; 5.

2. Divide a into two parts, such that $\tfrac{1}{m}$ of the first, plus $\tfrac{1}{n}$ of the second, shall be equal to b. Let $a=100$, $b=30$, $m=3$, $n=5$.

3. A sum of d dollars is divided between A and B. B receives b dollars as often as A receives a dollars. How much does each receive? Let $d=7000$, $a=3$, $b=2$.

4. If two trains start together and run in the same direction, one at the rate of m miles an hour, and the other at the rate of n miles an hour, after how many hours will they be d miles apart? Let $d=200$, $m=35$, $n=30$.

5. A farmer can plough a field in m days, and his son in n days; in how many days can they plough the field, working together? Let $m=10$, $n=15$.

6. A pupil was told to add m to a certain number, and to divide the sum by n. But he misunderstood the problem, and subtracted n from the number and multiplied the remainder by m. Nevertheless he obtained the correct result. What was the number? Let $m=12$, $n=13$.

7. One pipe can fill a cistern in a hours, a second in b hours, and a third in c hours. In how many hours can the three pipes fill the cistern, working together? Let $a=2$, $b=3$, $c=6$.

8. One pipe can fill a cistern in m hours, a second in n hours, and a third can empty it in p hours. After how many hours will the cistern be filled, if all pipes are open? Let $m=4$, $n=6$, $p=3$.

CHAPTER X.

SIMULTANEOUS LINEAR EQUATIONS.

SYSTEMS OF EQUATIONS.

1. If the linear equation in two unknown numbers
$$x+y=5 \tag{1}$$
be solved for y, we obtain
$$y=5-x.$$

We may substitute in this equation any particular numerical value for x and obtain a corresponding value for y. Thus, when $x=1$, $y=4$; when $x=2$, $y=3$; when $x=3$, $y=2$; etc.

In like manner the equation could have been solved for x in terms of y, and corresponding sets of values obtained.

Any set of corresponding values of x and y satisfies the given equation, and is therefore a solution.

2. Solving the equation
$$y-x=1 \tag{2}$$
for y, we have $y=1+x$. Then,

when $x=1$, $y=2$; when $x=2$, $y=3$; when $x=3$, $y=4$; etc.

Now, observe that equations (1) and (2) have the common solution, $x=2$, $y=3$. It seems evident, and it is proved in School Algebra, that these equations have only this solution in common.

Equations (1) and (2) express different relations between the unknown numbers, and are called **Independent Equations.**

Also, since they are satisfied by a common set of values of the unknown numbers, they are called **Consistent Equations.**

3. A System of Simultaneous Equations is a group of equations which are to be satisfied by the same set, or sets, of values of the unknown numbers.

A **Solution** of a system of simultaneous equations is a set of values of the unknown numbers which satisfies all of the equations.

4. *Two systems of equations are equivalent when every solution of either system is a solution of the other.*

E.g., the systems (I.) and (II.):

$$\left.\begin{array}{r}3x + 2y = 8, \\ x - y = 1,\end{array}\right\} \text{(I.)} \qquad \left.\begin{array}{r}3x + 2y = 8, \\ 2x - 2y = 2,\end{array}\right\} \text{(II.)}$$

are equivalent. For they are both satisfied by the solution, $x = 2$, $y = 1$, and, as we shall see later, by no other solution.

5. If the equations $\quad x + y = 7,$

$$x - y = 1,$$

be added, we obtain $\quad 2x = 8,$

in which the unknown number y does not appear. We say that y has been *eliminated* from the given equations.

6. Elimination is the process of deriving from two or more equations an equation which has one less unknown number.

Elimination by Addition and Subtraction.

7. Ex. 1. Solve the system $3x + 4y = 24,$ \hfill (1)

$$5x - 6y = 2. \hfill (2)$$

To eliminate y, we multiply the equations by such numbers as will make the coefficients of y numerically equal.

Multiplying (1) by 3, $\quad 9x + 12y = 72,$ \hfill (3)

Multiplying (2) by 2, $\quad 10x - 12y = 4,$ \hfill (4)

Adding (3) and (4), $\quad\quad 19x = 76.$

Whence $x = 4$.

Substituting 4 for x in (1), $12 + 4y = 24$.

Whence $y = 3$.

It is proved in School Algebra, Ch. XIII., that the above method is based upon equivalent equations.

Consequently the required solution is $x = 4$, $y = 3$.

This solution may be written 4, 3, it being understood that the first number is the value of x, and the second the value of y.

Ex. 2. Solve the system $12x + 15y = 8$. (1)

$16x + 9y = 7$. (2)

We will first eliminate x.

Multiplying (1) by 4, $48x + 60y = 32$. (3)

Multiplying (2) by 3, $48x + 27y = 21$. (4)

Subtracting (4) from (3), $33y = 11$.

Whence $y = \tfrac{1}{3}$.

Substituting $\tfrac{1}{3}$ for y in (1), $12x + 5 = 8$.

Whence $x = \tfrac{1}{4}$.

Consequently the required solution is $\tfrac{1}{4}$, $\tfrac{1}{3}$.

8. The examples of the preceding article illustrate the following method of elimination by addition and subtraction:

Multiply both members of the equations by such numbers as will make the coefficients of one of the unknown numbers numerically equal. Subtract, or add, corresponding members of the resulting equations, and equate the results.

Solve this equation in one unknown number. Substitute the value of this unknown number in the simpler of the given equations. Solve the resulting equation for the other unknown number.

The multipliers are obtained by dividing the L. C. M. of the coefficients of the unknown number to be eliminated by the coefficients of this unknown number. It is better to eliminate that unknown number which requires the smallest multipliers.

SYSTEMS OF EQUATIONS.

EXERCISES I.

Solve the following simultaneous equations:

1. $\begin{cases} x + y = 5, \\ x - y = 3. \end{cases}$
2. $\begin{cases} x + 2y = 7, \\ x - 2y = 4. \end{cases}$
3. $\begin{cases} x + 3y = 15, \\ 2x + y = 10. \end{cases}$
4. $\begin{cases} 3x + 2y = 26, \\ 5x - 4y = -8. \end{cases}$
5. $\begin{cases} 4x + 9y = -15, \\ 3x + 7y = -11. \end{cases}$
6. $\begin{cases} 5x + 8y = 19, \\ 2x + 3y = 8. \end{cases}$
7. $\begin{cases} 10x + 4y = 3, \\ 20y - 5x = 4. \end{cases}$
8. $\begin{cases} 5x + 2y = 11, \\ x - 3y = 9. \end{cases}$
9. $\begin{cases} 10x + 12y = 9, \\ 4x - 9y = -1. \end{cases}$
10. $\begin{cases} 8x + 6y = 1, \\ 12x - 14y = 13. \end{cases}$
11. $\begin{cases} 10x - 21y = 32, \\ 14y - 15x = -13. \end{cases}$
12. $\begin{cases} 3ax + 2by = c, \\ 3mx - 2ny = 0. \end{cases}$
13. $\begin{cases} \dfrac{x}{4} + \dfrac{y}{3} = 5, \\ \dfrac{x}{4} - \dfrac{y}{3} = -1. \end{cases}$
14. $\begin{cases} x - y = 1, \\ \dfrac{2}{5}x + \dfrac{3}{4}y = 5. \end{cases}$
15. $\begin{cases} \dfrac{x+y}{8} + \dfrac{x-y}{6} = 5, \\ \dfrac{x+y}{4} \cdot \dfrac{x-y}{3} = 10. \end{cases}$
16. $\begin{cases} 2x + \dfrac{y-2}{5} = 21, \\ 4y + \dfrac{x-4}{6} = 29. \end{cases}$
17. $\begin{cases} \dfrac{1}{x} + \dfrac{1}{y} = 7, \\ \dfrac{1}{x} - \dfrac{1}{y} = 5. \end{cases}$
18. $\begin{cases} \dfrac{4}{x} - 3y = 8, \\ \dfrac{5}{x} - 6y = 1. \end{cases}$
19. $\begin{cases} 7x - \dfrac{3}{y} = 16, \\ 3x - \dfrac{2}{y} = 4. \end{cases}$
20. $\begin{cases} \dfrac{1}{x} + \dfrac{1}{y} = \dfrac{1}{2}, \\ \dfrac{1}{x} - \dfrac{1}{y} = \dfrac{1}{6}. \end{cases}$

21. $\begin{cases} \dfrac{x+y}{4} - \dfrac{x-y}{5} = 0, \\ \dfrac{x+3y}{6} + \dfrac{x-3y}{3} = 5. \end{cases}$
22. $\begin{cases} \dfrac{7+x}{5} - \dfrac{2x-y}{4} = 3y-5, \\ \dfrac{5y-7}{2} + \dfrac{4x-3}{6} = 18-5x. \end{cases}$

23. $\begin{cases} \dfrac{x}{m-a} + \dfrac{y}{m-b} = 1, \\ \dfrac{x}{n-a} + \dfrac{y}{n-b} = 1. \end{cases}$
24. $\begin{cases} \dfrac{x}{n+1} + \dfrac{y}{n-1} = \dfrac{1}{n-1}, \\ \dfrac{x}{n-1} + \dfrac{y}{n+1} = \dfrac{1}{n^2-1}. \end{cases}$

Elimination by Comparison.

9. Ex. Solve the system
$$7x + 2y = 20, \qquad (1)$$
$$13x - 3y = 17. \qquad (2)$$

To eliminate y, we proceed as follows:

Solving (1) for y, $\qquad y = \dfrac{20 - 7x}{2}. \qquad (3)$

Solving (2) for y, $\qquad y = \dfrac{13x - 17}{3}. \qquad (4)$

Equating these values of y,
$$\dfrac{20 - 7x}{2} = \dfrac{13x - 17}{3}. \qquad (5)$$

Whence $\qquad x = 2.$

Substituting 2 for x in (3), $y = \dfrac{20 - 14}{2} = 3.$

It is proved in School Algebra, Ch. XIII., that the above method is based upon equivalent equations.

Consequently the required solution is 2, 3.

10. This example illustrates the following method of elimination by comparison:

Solve the given equations for the unknown number to be eliminated, and equate the expressions thus obtained. The derived equation will contain but one unknown number.

Solve this derived equation, and substitute the value of the unknown number thus obtained in the simplest of the preceding equations. Solve the resulting equation for the other unknown number.

EXERCISES II.

Solve the following simultaneous equations:

1. $\begin{cases} x = 5y + 7, \\ x = 3y + 11. \end{cases}$
2. $\begin{cases} y = 4x - 2, \\ y = 3x + 1. \end{cases}$
3. $\begin{cases} 3x = 5y + 2, \\ 4x = 3y + 10. \end{cases}$
4. $\begin{cases} 4y = x + 1, \\ 2y = x - 1. \end{cases}$
5. $\begin{cases} 3x - 2 = 7y - 6, \\ 2x + 8 = 6y. \end{cases}$
6. $\begin{cases} 4x - 7y = 25, \\ 3y - 2x = -11. \end{cases}$
7. $\begin{cases} 6x - 7y = 47, \\ 4x + 3y = -7. \end{cases}$
8. $\begin{cases} 8y + x = -25, \\ y + 8x = -11. \end{cases}$
9. $\begin{cases} \dfrac{x}{6} + \dfrac{y}{2} = 3, \\ \dfrac{x}{3} - \dfrac{y}{5} = \dfrac{6}{5}. \end{cases}$
10. $\begin{cases} \dfrac{x}{7} + 7y = 99, \\ \dfrac{y}{7} + 7x = 51. \end{cases}$
11. $\begin{cases} \dfrac{x+y}{x-y} = \dfrac{7}{3}, \\ \dfrac{x+1}{y+1} = \dfrac{12}{5}. \end{cases}$
12. $\begin{cases} \dfrac{3x+7}{10y+3} = 1, \\ \dfrac{12x+5}{7y+1} = 2. \end{cases}$

Elimination by Substitution.

11. Ex. Solve the system

$$5x - 2y = 1, \qquad (1)$$
$$4x + 5y = 47. \qquad (2)$$

If we wish to eliminate x, we proceed as follows:

Solving (1) for x, $\qquad x = \dfrac{1 + 2y}{5}. \qquad (3)$

Substituting $\dfrac{1+2y}{5}$ for x in (2),

$$4\left(\dfrac{1+2y}{5}\right) + 5y = 47. \qquad (4)$$

Whence $\qquad y = 7.$

Substituting 7 for y in (3), $x = 3$.

It is proved in School Algebra, Ch. XIII., that the above method is based upon equivalent equations.

Consequently the required solution is 3, 7.

12. This example illustrates the following method of elimination by substitution:

Solve the simpler equation for the unknown number to be eliminated in terms of the other. Substitute the value thus obtained in the other equation. The derived equation will contain but one unknown number.

Solve the derived equation, and substitute the value of the unknown number thus obtained in the expression for the other unknown number. Solve the resulting equation.

EXERCISES III.

1. $\begin{cases} 3y - x = -15, \\ x = 8y. \end{cases}$
2. $\begin{cases} y - 6x = 4, \\ y = 14x. \end{cases}$

3. $\begin{cases} 15y - 3x = 21, \\ 3y = 2x. \end{cases}$
4. $\begin{cases} x = y + 5, \\ 3x - y = 19. \end{cases}$

5. $\begin{cases} 5x + 2y = 11, \\ 3x - 5y = -12. \end{cases}$
6. $\begin{cases} 4y + 5x = 31, \\ 3x - 2y = 1. \end{cases}$

7. $\begin{cases} \dfrac{x+1}{y} = \dfrac{1}{4}, \\ \dfrac{x}{y+1} = \dfrac{1}{5}. \end{cases}$
8. $\begin{cases} \dfrac{x}{3} + 4y = 66, \\ 3x + \dfrac{y}{4} = 22. \end{cases}$

9. $\begin{cases} \dfrac{2x-1}{2y+1} = 1, \\ x + y = 15. \end{cases}$
10. $\begin{cases} \dfrac{3x - 2y}{3x - y} = \dfrac{5}{8}, \\ x + y = 20. \end{cases}$

11. $\begin{cases} \dfrac{3x+4}{3y+5} = 2, \\ \dfrac{4x+3}{3y+2} = 3. \end{cases}$
12. $\begin{cases} \dfrac{1}{2x-3y} = \dfrac{9}{3x-2y}, \\ \dfrac{1}{2x-9} = \dfrac{9}{4y-3}. \end{cases}$

Linear Equations in Three Unknown Numbers.

13. The following example will illustrate methods of solving systems of three linear equations in three unknown numbers:

Ex. Solve the system

$$2x - 3y + 5z = 11, \quad (1)$$
$$5x + 4y - 6z = -5, \quad (2)$$
$$-4x + 7y - 8z = -14. \quad (3)$$

To eliminate x, we proceed as follows:

Multiplying (1) by 5, $\quad 10x - 15y + 25z = 55.$ (4)

Multiplying (2) by 2, $\quad 10x + 8y - 12z = -10.$ (5)

Subtracting (4) from (5), $\quad 23y - 37z = -65.$ (6)

Multiplying (1) by 2, $\quad 4x - 6y + 10z = 22.$ (7)

Adding (3) and (7), $\quad y + 2z = 8.$ (8)

Solving (6) and (8), $\quad y = 2.$

$$z = 3.$$

Substituting 2 for y and 3 for z in (1), $\quad x = 1.$

Consequently the required solution is 1, 2, 3.

14. This example illustrates the following method:

Eliminate one of the unknown numbers from any two of the equations; next eliminate the same unknown number from the third equation and either of the other two. Two equations in the same two unknown numbers are thus derived.

Solve these equations for the two unknown numbers, and substitute the values thus obtained in the simplest equation which contains the third unknown number.

EXERCISES IV.

1. $\begin{cases} x+y=64, \\ x+z=54, \\ y+z=44. \end{cases}$
2. $\begin{cases} x+y=9; \\ x-y=3, \\ x-z=4, \end{cases}$
3. $\begin{cases} x+2y=15, \\ y+2z=10, \\ z+2x=17. \end{cases}$

4. $\begin{cases} 3x+y=18, \\ 3y+z=10, \\ 3z+x=8. \end{cases}$
5. $\begin{cases} x=3y-13, \\ y=8z-17, \\ z=4x-29. \end{cases}$
6. $\begin{cases} 2x=y+6, \\ 3y=z+9, \\ 4z=x+7. \end{cases}$

7. $\begin{cases} 5x-y=33, \\ 2x+z=21, \\ 3y-z=16. \end{cases}$
8. $\begin{cases} 4x-3y=13, \\ 3x-4z=13, \\ 5y-6z=13. \end{cases}$

9. $\begin{cases} 3x+25y-6z=35, \\ 6x-10y+21z=49, \\ 8x+15y-14z=-4. \end{cases}$
10. $\begin{cases} x+y+z=14, \\ 6x-3y-7z=0, \\ 4x-9y+7z=0. \end{cases}$

11. $\begin{cases} 8x+21y-9z=-23, \\ 12x-28y-15z=89, \\ 15x-49y+18z=29. \end{cases}$
12. $\begin{cases} 2x+3y-4z=1, \\ 3x+4z-5y=2, \\ 4y+5z-6x=3. \end{cases}$

13. $\begin{cases} \dfrac{x}{3}+\dfrac{y}{4}=6, \\ \dfrac{y}{2}+\dfrac{z}{3}=6, \\ \dfrac{x}{4}+\dfrac{z}{2}=6. \end{cases}$
14. $\begin{cases} \dfrac{1}{x}+\dfrac{1}{y}=5, \\ \dfrac{1}{x}+\dfrac{1}{z}=6, \\ \dfrac{1}{y}+\dfrac{1}{z}=7. \end{cases}$

Problems.

15. Pr. 1. The sum of the two digits of a number is 12. If the digits are interchanged, the resulting number will exceed the original one by three-fourths of the original number. What is the number?

Let x stand for the units' digit, and y for the tens' digit.
Then the original number is $10y+x$.
When the digits are interchanged, the resulting number is $10x+y$.

The first condition of the problem states,

in *verbal* language: *the sum of the digits is* 12;

in *algebraic* language: $\quad x + y = 12.$ \hfill (1)

The second condition states,

in *verbal* language: *the resulting number minus the original number is equal to ¾ of the original number;*

in *algebraic* language: $10x + y - (10y + x) = \frac{3}{4}(10y + x).$ \hfill (2)

Solving (1) and (2), $\quad x = 8, \ y = 4.$

Therefore the required number is 48.

Pr. 2. A tank can be filled by two pipes. If the first is left open 6 minutes, and the second 7 minutes, the tank will be filled; or if the first is left open 3 minutes, and the second 12 minutes, the tank will be filled. In what time can each pipe fill the tank?

Let x stand for the number of minutes it takes the first pipe to fill the tank, and y for the number of minutes it takes the second pipe. Let the capacity of the tank be represented by 1. Then in 1 minute the first pipe fills $\frac{1}{x}$ of the tank, and in 6 minutes $\frac{6}{x}$ of the tank; the second pipe fills $\frac{7}{y}$ of the tank in 7 minutes. Therefore, by the conditions of the problem,

$$\frac{6}{x} + \frac{7}{y} = 1; \quad \frac{3}{x} + \frac{12}{y} = 1.$$

Whence $\quad x = 10\frac{1}{5}, \ y = 17.$

EXERCISES V.

1. The sum of two numbers is 18; their quotient is 5. What are the numbers?

2. The sum of the sides of a rectangular field is 100 yards. The difference of two adjacent sides is 10 yards. What are the dimensions of the field?

3. A man bought iron and steel at different times for the same prices. The first time he bought 125 pounds of iron and

80 pounds of steel for $9. The second time, 150 pounds of iron and 120 pounds of steel for $12. What was the cost of each per pound?

4. An engineer was laying a single rail 800 feet in length with short and long rails which he had no means of cutting. He found that 40 of the short and 8 of the long, or 30 of the short and 16 of the long, would answer his purpose. What were the lengths of the rails?

5. A contractor had to move 210 cubic yards of earth twice. The first time he had 20 cart loads and 30 wagon loads. The second time he had 15 cart loads and 33 wagon loads. How many yards in a cart load and in a wagon load?

6. If 5 is added to the numerator of a certain fraction, and 6 is subtracted from its denominator, the value of the resulting fraction is $\frac{7}{4}$. But if 6 is added to the numerator and 5 is subtracted from the denominator, the value of the resulting fraction is $\frac{8}{10}$. What is the fraction?

7. If the numerator of a certain fraction is multiplied by 2 and the denominator increased by 7, the value of the resulting fraction will be $\frac{2}{3}$; but if the numerator is multiplied by 3 and the denominator increased by 1, the value of the resulting fraction is $\frac{3}{2}$. What is the fraction?

8. A man made a journey, 8 hours by boat and 6 hours by rail. Coming home, he rode 13 hours by boat and 4 hours by rail. Altogether he travelled 552 miles. How many miles per hour did the boat and the train make?

9. The sum of the two digits of a number is 7. If 9 is subtracted from the number, the digits will be interchanged. What is the number?

10. The sum of the two digits of a number is 9. If the number is divided by 12, the quotient will be the tens' digit. What is the number?

11. A sum of money at simple interest amounted in 6 years to $1700, and in 10 years to $2000. What was the principal and rate?

12. The sum of two numbers is 47. If the greater be divided by the less, the quotient and the remainder will each be 5. What are the numbers?

13. A father said to his son: "In 5 years I shall be four times as old as you are now, but in 2 years I shall be three times as old as you will be then." How old was the father and the son?

14. A father is now 27 years older than his son. Ten years ago he was ten times as old as his son. What are their ages?

15. A man had a number of pieces of iron of two different weights. He wished to get the weight of each piece, but had only a 5-pound weight. He found that 6 of the first size and 1 of the second would just balance the 5-pound weight; also that 2 of the first and 4 of the second would also balance the weight. What was the weight of each sample?

16. Two workmen do a piece of work together. If A works 12 days and B 8 days, the work will be completed; but if A works 9 days and B 12 days, it will also be completed. In how many days can each do the work alone? In how many days, working together?

17. Two numbers are in the ratio of 3 to 5. If the first is increased by 10, and the second decreased by 10, they are in the ratio of 5 to 3. What are the numbers?

Note. — If x and y are in the ratio of a to b, then $\dfrac{x}{y} = \dfrac{a}{b}$.

18. A painter has two kinds of white paint, the prices per gallon being in a certain ratio. If he mixes 3 gallons of the first and 2 of the second, the mixture will be worth 95 cents per gallon. A mixture of 4 gallons of the first and 6 gallons of the second will be worth $1.05 per gallon. What was the cost of each kind per gallon and the ratio?

19. A farmer took oats and corn to market. Twenty bushels of oats and 30 bushels of corn became mixed, and he sold the mixture at 31 cents per bushel. He then made another mixture of 30 bushels of oats and 20 bushels of corn, and sold it at 34 cents per bushel. On both sales he cleared the same amount that the oats and corn would have brought if sold separately. What was the price of the oats and corn per bushel?

20. A owes $75 and B owes $90, but neither has enough to pay his debt. A could pay his debt if he had $\frac{1}{5}$ of B's money; B could pay his debt if he had $\frac{3}{10}$ of A's money. How much had each?

21. A man has a horse and carriage with harness. The harness is worth $25. The cost of the horse and harness is worth $2\frac{1}{2}$ times the cost of the carriage. The cost of the carriage and harness is worth $\frac{3}{4}$ of the cost of the horse. What is the cost of each?

22. A cistern has an outlet and an inlet controlled by valves. If both are wide open, the cistern will contain 600 gallons after 8 hours. If, however, the outlet is $\frac{2}{3}$ open, and the inlet is $\frac{1}{4}$ shut, the cistern will contain 440 gallons after 8 hours. How many gallons can pass through each pipe per hour?

23. The sum of three numbers is 49. The first minus the second gives a remainder 5; the second minus the third gives a remainder 4. What are the numbers?

24. The sums of the sides of a triangle, taken in pairs, are 11, 10, and 15. What are the lengths of the sides?

25. The joint ages of a father and his two sons are 44 years. The father's age now is three times the joint ages of his sons 2 years ago. Three years hence the father's age will be three times that of his oldest son. What are their ages?

26. The sum of 3 digits of a number is 10. If 297 be added to the number, the digits are reversed. If 360 be added to the number, the digits in the hundreds' and tens' places are interchanged. What is the number?

27. A tank can be filled by pipes A and B in 35 minutes, A and C in 32 minutes, and B and C in 70 minutes. How long will it take each pipe to fill the tank alone?

CHAPTER XI.

INVOLUTION.

1. Involution is the process of raising a number to any required power.

Powers of Powers.

2. Ex. $(a^4)^5 = a^4 a^4 a^4 a^4 a^4 = a^{4+4+4+4+4} = a^{4\times 5} = a^{20}$.

This example illustrates the following method of finding any required power of a given power:

Multiply the exponent of the given power by the exponent of the required power; or, stated symbolically,

$$(a^m)^n = a^{mn}.$$

Powers of Products.

3. Ex. $(ab)^4 = (ab)(ab)(ab)(ab)$
$= (aaaa)(bbbb) = a^4 b^4.$

This example illustrates the following method of finding any required power of a product:

Take the product of the factors, each raised to the required power; or, stated symbolically,

$$(ab)^n = a^n b^n; \quad (abc)^n = a^n b^n c^n; \text{ etc.}$$

4. The principles of Arts. 2–3 give the following method of raising a monomial to any required power.

Raise the numerical coefficient to the required power, and multiply the exponent of each literal factor by the exponent of the required power.

Ex. $(4 a^3 b)^2 = 4^2 a^{3\times 2} b^2 = 16\, a^6 b^2.$

Powers of Fractions.

5. Ex. 1. $\left(\dfrac{2}{5}\right)^3 = \dfrac{2}{5} \times \dfrac{2}{5} \times \dfrac{2}{5} = \dfrac{2 \times 2 \times 2}{5 \times 5 \times 5} = \dfrac{2^3}{5^3} = \dfrac{8}{125}.$

This example illustrates the following method of raising a fraction to a power:

Raise each term of the fraction to the required power; or, stated symbolically,

$$\left(\dfrac{a}{b}\right)^n = \dfrac{a^n}{b^n}.$$

Ex. 2. $\left(\dfrac{2\,a^2b^3}{c^4}\right)^3 = \dfrac{2^3(a^2)^3(b^3)^3}{(c^4)^3} = \dfrac{8\,a^6b^9}{c^{12}}.$

EXERCISES I.

Write the squares, the cubes, and the fourth powers of:

1. x^2.
2. $-x^4$.
3. $2\,x^7$.
4. $-3\,ab$.
5. $5\,ab^2$.
6. $4\,x^2y^3$.
7. $2\,m^2xy^5$.
8. $5\,a^2b^5c^6$.
9. $\dfrac{a}{b}$
10. $\dfrac{2\,a}{y}$.
11. $-\dfrac{3\,x^2}{2\,y^3}$.
12. $-\dfrac{4\,x^2y}{3\,ab^3}$.

Find the values of each of the following powers:

13. $(-3\,xy^4)^2$.
14. $(4\,abc^3)^2$.
15. $(2\,a^2b^3)^3$.
16. $(-3\,x^2y^4)^3$.
17. $(5\,a^5b^6c)^2$.
18. $(-4\,x^4y^2z^5)^3$.
19. $(2\,xy^2z^3)^4$.
20. $(-a^2xy^4)^4$.
21. $(-2\,m^2n^3)^5$.
22. $(ab^4c^2d^3)^5$.
23. $(-ax^2y^3)^6$.
24. $(3\,x^3yz^4)^6$.
25. $(2\,m^2x^7y^4)^7$.
26. $(-a^2b^5c^2)^7$.
27. $(-x^2y^3z^4)^9$.
28. $\left(\dfrac{ab^2c}{5\,xy^3}\right)^2$.
29. $\left(-\dfrac{6\,a^4x^6}{7\,b^3y^5}\right)^2$.
30. $\left(\dfrac{3\,ab^3c^5}{x^2y^4z^6}\right)^2$.
31. $\left(\dfrac{3\,a^2b}{4\,c^2d^2}\right)^3$.
32. $\left(-\dfrac{3\,a^2b^5}{4\,m^2n^3}\right)^3$.
33. $\left(-\dfrac{a^2bc^3}{2\,xy^2z}\right)^4$.

The Cube of a Binomial.

6. By actual multiplication, we obtain

$$(a+b)^3 = (a^2+2ab+b^2)(a+b) = a^3+3a^2b+3ab^2+b^3, \quad (1)$$

and $(a-b)^3 = (a^2-2ab+b^2)(a-b) = a^3-3a^2b+3ab^2-b^3. \quad (2)$

Ex. 1. $(5x+3y)^3 = (5x)^3 + 3(5x)^2(3y) + 3(5x)(3y)^2 + (3y)^3$
$= 125x^3 + 225x^2y + 135xy^2 + 27y^3.$

Ex. 2. $(2a-b)^3 = (2a)^3 - 3(2a)^2b + 3(2a)b^2 - b^3$
$= 8a^3 - 12a^2b + 6ab^2 - b^3.$

Fourth Power of a Binomial.

7. By actual multiplication, we obtain

$$(a+b)^4 = (a^2+2ab+b^2)(a^2+2ab+b^2)$$
$$= a^4 + 4a^3b + 6a^2b^2 + 4ab^3 + b^4.$$

$$(a-b)^4 = a^4 - 4a^3b + 6a^2b^2 - 4ab^3 + b^4.$$

Ex. $(2a+3b)^4 = (2a)^4 + 4(2a)^3(3b) + 6(2a)^2(3b)^2$
$+ 4(2a)(3b)^3 + (3b)^4$
$= 16a^4 + 96a^3b + 216a^2b^2 + 216ab^3 + 81b^4.$

EXERCISES II.

Raise each of the following expressions to the required power:

1. $(a+1)^3.$
2. $(a-2)^3.$
3. $(2x+3)^3.$
4. $(3-4y)^3.$
5. $(a+2b)^3.$
6. $(3a-b)^3.$
7. $(2x-3y)^3.$
8. $(ax+by)^3.$
9. $(x^2+2x)^3.$
10. $(1+x)^4.$
11. $(x-1)^4.$
12. $(2a+b)^4.$
13. $(x-3y)^4.$
14. $(2x+4y)^4.$
15. $(3x^2-2x)^4$

CHAPTER XII.

EVOLUTION.

1. A **Root** of a number is one of the equal factors of the number.

E.g., 2 is a root of 4, of 8, of 16, etc.

2. A **Second**, or **Square Root** of a number is one of *two* equal factors of the number.

E.g., since $5 \times 5 = 25$ and $(-5)(-5) = 25$, therefore $+5$ and -5 are square roots of 25.

A **Third**, or **Cube Root** of a number is one of *three* equal factors of the number.

E.g., since $3 \times 3 \times 3 = 27$, therefore 3 is a cube root of 27; since $(-3)(-3)(-3) = -27$, therefore -3 is a cube root of -27.

In general, the *q*th *root* of a number is one of q equal factors of the number.

E.g., a qth root of x^q is x.

3. The **Radical Sign**, $\sqrt{}$, is used to denote a root, and is placed before the number whose root is to be found.

The **Radicand** is the number whose root is required.

The **Index** of a root is the number which indicates what root is to be found, and is written over the radical sign. The index 2 is usually omitted.

E.g., $\sqrt[2]{9}$, or $\sqrt{9}$, denotes a second, or square root of 9; the radicand is 9, and the index is 2.

4. A vinculum is often used in connection with the radical sign to indicate what part of an expression following the sign is affected by it.

E.g., $\sqrt{9} + 16$ means the sum of $\sqrt{9}$ and 16, while $\sqrt{9+16}$ means a square root of the sum $9 + 16$. Likewise $\sqrt[3]{a^3} \times b^6$ means the product of $\sqrt[3]{a^3}$ and b^6, while $\sqrt[3]{a^3 \times b^6}$ means a cube root of $a^3 b^6$.

Parentheses may be used instead of the vinculum in connection with the radical sign; as $\sqrt{(9+16)}$ for $\sqrt{9+16}$.

5. It follows from the definition of a root that the square of the square root of a number is the number, the cube of the cube root of a number is the number, and so on.

E.g., $(\sqrt{4})^2 = 4$; $(\sqrt[3]{8})^3 = 8$; etc.

6. An **Even Root** is one whose *index* is *even;* as $\sqrt{a^2}$, $\sqrt[4]{a^4}$, $\sqrt[2q]{a^{2q}}$

An **Odd Root** is one whose *index* is *odd;* as $\sqrt[3]{8}$, $\sqrt[5]{a^{10}}$, $\sqrt[2q+1]{a^{2q+1}}$

Number of Roots.

7. Since $(\pm 4)^2 = 16$, therefore $\sqrt{16} = \pm 4$;

since $(\pm a)^4 = a^4$, therefore $\sqrt[4]{a^4} = \pm a$.

These examples illustrate the principle:

A positive number has at least two even roots, equal and opposite; i.e., *one positive and one negative.*

8. Since $(-3)^3 = -27$, therefore $\sqrt[3]{-27} = -3$;

since $2^5 = 32$, therefore $\sqrt[5]{32} = 2$.

These examples illustrate the principle:

A positive or a negative number has at least one odd root of the same sign as the number itself.

9. It is proved in School Algebra, Ch. XVIII., that $\sqrt{2}$ cannot be expressed either as an integer or as a fraction. That is, there is no integer or fraction whose square is 2. Such numbers as $\sqrt{2}$, $\sqrt[3]{a^2}$, etc., are called **Irrational Numbers.**

It is there proved that the value of an irrational number can be found approximately to any degree of accuracy.

E.g., approximate values of $\sqrt{2}$ are 1.4, 1.41, 1.414, etc.

The numbers hitherto used in this book, that is, integers and fractions, are called **Rational Numbers.**

10. *Since* $(+4)^2 = +16$ *and* $(-4)^2 = +16$, *there is no number, with which we are as yet familiar, whose square is* -16.

Consequently $\sqrt{-16}$ cannot be expressed in terms of the numbers as yet used in this book.

Such roots are called **Imaginary Numbers**.

For the sake of distinction both rational and irrational numbers are called **Real Numbers**.

Evolution.

11. **Evolution** is the process of finding a root of a given number.

12. *In the following articles the radicands are limited to positive values, and the roots to positive roots.*

13. Since $(a^2)^3 = a^6$, therefore $\sqrt[3]{a^6} = a^2 = a^{\frac{6}{3}}$.

This example illustrates the principle:

(i.) *The root of a power is obtained by dividing the exponent of the power by the index of the root.*

E.g., $\qquad \sqrt[5]{a^{15}} = a^{\frac{15}{5}} = a^3$.

Since $(ab)^2 = a^2b^2$, therefore $\sqrt{(a^2b^2)} = ab = \sqrt{a^2} \times \sqrt{b^2}$.

This example illustrates the principle:

(ii.) *The root of a product of two or more factors is equal to the product of the like roots of the factors, and conversely.*

E.g., $\quad \sqrt{(16 \times 25)} = \sqrt{16} \times \sqrt{25} = 4 \times 5 = 20$;

$\qquad \sqrt[3]{(8\,a^3b^6)} = \sqrt[3]{8} \times \sqrt[3]{a^3} \times \sqrt[3]{b^6} = 2 \times a \times b^2 = 2\,ab^2$

Since $\left(\dfrac{a}{b}\right)^2 = \dfrac{a^2}{b^2}$, therefore $\sqrt{\dfrac{a^2}{b^2}} = \dfrac{a}{b} = \dfrac{\sqrt{a^2}}{\sqrt{b^2}}$.

This example illustrates the principle:

(iii.) *The root of a quotient of two numbers is equal to the quotient of the like roots of the numbers, and conversely.*

E.g., $\quad \sqrt{\dfrac{25}{16}} = \dfrac{\sqrt{25}}{\sqrt{16}} = \dfrac{5}{4}$; $\quad \sqrt[3]{\dfrac{27\,a^3}{b^6}} = \dfrac{\sqrt[3]{(27\,a^3)}}{\sqrt[3]{b^6}} = \dfrac{3\,a}{b^2}$.

Roots of Monomials.

14. The *positive* root of a positive number can be found by applying the principles of Art. 13.

The *negative even* root of a positive number is found by prefixing the negative sign to its positive root.

Since $\sqrt[3]{-8} = -2$, and $-\sqrt[3]{8} = -2$,

therefore $\sqrt[3]{-8} = -\sqrt[3]{8}.$

That is, the *negative odd* root of a negative number is found by prefixing the negative sign to the positive root of the radicand taken positively.

Ex. 1. $\sqrt{(16\,a^2b^4)} = \sqrt{16} \times \sqrt{a^2} \times \sqrt{b^4}$
$= 4\,ab^2$, the positive square root.

Therefore $\pm\sqrt{(16\,a^2b^4)} = \pm 4\,ab^2.$

In the following examples we shall give only the *positive* even roots.

Ex. 2. $\sqrt[3]{(-27\,x^3y^6z^9)} = \sqrt[3]{-27} \times \sqrt[3]{x^3} \times \sqrt[3]{y^6} \times \sqrt[3]{z^9}$
$= -3\,xy^2z^3.$

Ex. 3. $\sqrt[4]{\dfrac{16\,a^8b^{12}}{625\,c^{16}}} = \dfrac{\sqrt[4]{(16\,a^8b^{12})}}{\sqrt[4]{(625\,c^{16})}} = \dfrac{\sqrt[4]{16} \times \sqrt[4]{a^8} \times \sqrt[4]{b^{12}}}{\sqrt[4]{625} \times \sqrt[4]{c^{16}}} = \dfrac{2\,a^2b^3}{5\,c^4}.$

EXERCISES I.

Find the values of the following expressions:

1. $\sqrt{x^4}.$
2. $\sqrt{a^8}.$
3. $\sqrt[3]{z^9}.$
4. $\sqrt[3]{-x^6}.$
5. $\sqrt[3]{(8\,y^9)}.$
6. $\sqrt{(9\,x^2y^2)}.$
7. $\sqrt[3]{(-64\,z^{12})}.$
8. $\sqrt[3]{(8\,x^3y^3)}.$
9. $\sqrt[3]{(-8\,m^6n^6)}.$
10. $\sqrt{(36\,a^2b^4c^6)}.$
11. $\sqrt[5]{(a^{10}b^{15}c^5)}.$
12. $\sqrt[7]{(m^{21}n^{35})}.$
13. $\sqrt[4]{(16\,a^4b^8c^{12})}.$
14. $\sqrt{\dfrac{a^4}{x^6}}.$
15. $\sqrt{\dfrac{m^2n^2}{x^4y^6}}.$
16. $\sqrt[3]{\dfrac{x^6y^9}{8\,m^{12}}}.$
17. $\sqrt[4]{\dfrac{256\,a^8}{81\,b^4c^{12}}}.$
18. $\sqrt[5]{-\dfrac{32\,x^{10}y^{15}}{b^5c^{10}}}.$
19. $\sqrt[3]{-\dfrac{8\,a^6b^9c^6}{27\,x^3y^6z^9}}.$

ROOTS OF MULTINOMIALS.

Square Roots of Multinomials.

15. The square root of a trinomial which is the square of a binomial can be found by inspection (Ch. VI., Art. 9).

16. Since $(a + b)^2 = a^2 + 2ab + b^2$,

we have $\sqrt{(a^2 + 2ab + b^2)} = a + b$.

From this identity we infer:

(i.) *The first term of the root is the square root of the first term of the trinomial;* i.e., $a = \sqrt{a^2}$.

(ii.) *If the square of the first term of the root be subtracted from the trinomial, the remainder will be*

$$2ab + b^2, \;= (2a + b)b.$$

Twice the first term of the root, $2a$, is called the **Trial Divisor**.

(iii.) *The second term of the root is obtained by dividing the first term of the remainder by the trial divisor;* i.e., $b = \dfrac{2ab}{2a}$.

The trial divisor plus the second term of the root is called the **Complete Divisor**.

(iv.) *If the product of the complete divisor by the second term of the root be subtracted from the first remainder, the second remainder will be* 0.

The work may be arranged as follows:

$$\begin{array}{r|ll}
a^2 + 2ab + b^2 & a + b & \\
a^2 & 2a & \text{trial divisor} \\
\hline
2ab & 2ab \div 2a = b, & \text{second term of root} \\
& 2a + b & \text{complete divisor} \\
2ab + b^2 & = (2a + b)b &
\end{array}$$

17. Ex. 1. Find the square root of $4x^4 - 12x^2y + 9y^2$.

The work, arranged as above, writing only the trial and the complete divisor, is:

$$\begin{array}{r|l} 4x^4 - 12x^2y + 9y^2 & 2x^2 - 3y \\ 4x^4 & 4x^2 \\ \hline -12x^2y & \\ -12x^2y + 9y^2 & 4x^2 - 3y \end{array}$$

The square root of $4x^4$ is $2x^2$, the first term of the root. The trial divisor is $2(2x^2), = 4x^2$. The second term of the root is $-\dfrac{12x^2y}{4x^2}, = -3y$. The complete divisor is $4x^2 - 3y$.

Ex. 2. Find the square root of
$$4x^4 - 12x^3 + 29x^2 - 30x + 25.$$

The work follows:

$$\begin{array}{r|l} 4x^4 - 12x^3 + 29x^2 - 30x + 25 & 2x^2 - 3x + 5 \\ 4x^4 & 4x^2 \\ \hline -12x^3 & \\ -12x^3 + 9x^2 & 4x^2 - 3x \\ \hline 20x^2 & \\ 20x^2 - 30x + 25 & 4x^2 - 6x + 5 \end{array}$$

Only the trial divisor and the complete divisor of each stage are written, the other steps being performed mentally.

The square root of $4x^4$ is $2x^2$, the first term of the root. The trial divisor is $2(2x^2), = 4x^2$. The second term of the root is $-\dfrac{12x^3}{4x^2}, = -3x$. The complete divisor is $4x^2 - 3x$, which is multiplied by the second term of the root, giving $-12x^3 + 9x^2$. The first term of the second remainder is $20x^2$.

The third term of the root is $\dfrac{20x^2}{4x^2}, = 5$.

To form the complete divisor at this stage, we multiply the part of the root previously found, $2x^2 - 3x$, by 2, and to the product add the term just found. We thus obtain $4x^2 - 6x + 5$. This complete divisor we multiply by the last term of the root.

18. The preceding method can be extended to find square roots which are multinomials of any number of terms.

The work consists of repetitions of the following steps:

After one or more terms of the root have been found, obtain each succeeding term, by dividing the first term of the remainder at that stage by twice the first term of the root.

Find the next remainder by subtracting from the last remainder the expression $(2a + b)b$, wherein a stands for the part of the root already found, and b for the term last found.

EXERCISES II.

Find the square roots of the following:

1. $x^2 + 6x + 9$.
2. $m^2 + 4mn + 4n^2$.
3. $4a^2 + 49b^2 - 28ab$.
4. $25a^2 - 10a + 1$.
5. $25x^2 + 9y^2 - 30xy$.
6. $a^2 + b^2 + 2ab + 2a + 2b + 1$.
7. $x^2 + 9y^2 + 9 - 6xy + 6x - 18y$.
8. $4a^4 + b^2 + 1 - 4a^2b + 4a^2 - 2b$.
9. $4a^4 + b^2 + a^2 - 4a^2b + 4a^3 - 2ab$.
10. $a^4 + 3a^2 + 1 + 2a^3 + 2a$.
11. $m^4 + 13m^2 + 4 - 6m^3 - 12m$.
12. $a^6 + a^4 + b^2 + 2a^5 - 2a^3b - 2a^2b$.
13. $9m^4 + 10m^2 + 12m^3 + 4m + 1$.
14. $x^4 + 4y^4 + 9z^4 + 4x^2y^2 + 6x^2z^2 + 12y^2z^2$.
15. $a^2 + 4a^2b^2 + 9b^2 + 4a^2b + 6ab + 12ab^2$.

Cube Roots of Multinomials.

19. The process of finding the cube root of a multinomial is the inverse of the process of cubing the multinomial.

Since
$$(a + b)^3 = a^3 + 3a^2b + 3ab^2 + b^3$$
$$= a^3 + (3a^2 + 3ab + b^2)b, \qquad (1)$$
we have $\quad \sqrt[3]{(a^3 + 3a^2b + 3ab^2 + b^3)} = a + b. \qquad (2)$

ROOTS OF MULTINOMIALS.

From the identity (2), we infer:

(i.) *The first term of the root is the cube root of the first term of the multinomial;* i.e., $a = \sqrt[3]{a^3}$.

(ii.) *If the cube of the first term of the root be subtracted from the multinomial, the remainder will be*

$$3a^2b + 3ab^2 + b^3, = (3a^2 + 3ab + b^2)b.$$

Three times the square of the first term of the root, $3a^2$, is called the **Trial Divisor**.

(iii.) *The second term of the root is obtained by dividing the first term of the remainder by the trial divisor;* i.e., $b = \dfrac{3a^2b}{3a^2}$.

The sum $3a^2 + 3ab + b^2$ is called the **Complete Divisor**.

(iv.) *If the product of the complete divisor by the second term of the root be subtracted from the first remainder, the second remainder will be 0.*

The work may be arranged as follows:

$$
\begin{array}{l|ll}
a^3 + 3a^2b + 3ab^2 + b^3 & a+b & \\
a^3 & 3a^2 & \text{trial divisor} \quad (1) \\
\cline{1-1}
3a^2b & 3a^2b \div 3a^2 = b & \text{second term of root} \quad (2) \\
 & 3a^2 + 3ab + b^2, & \text{complete divisor} \quad (3) \\
3a^2b + 3ab^2 + b^3 & = (3a^2 + 3ab + b^2) \times b & \quad (4)
\end{array}
$$

20. Ex. 1. Find the cube root of $27x^3 + 54x^2y + 36xy^2 + 8y^3$

The work, arranged as above, is:

$$
\begin{array}{l|l}
27x^3 + 54x^2y + 36xy^2 + 8y^3 & 3x + 2y \\
27x^3 & 3(3x)^2 = 27x^2 \\
\cline{1-1}
54x^2y & \\
54x^2y + 36xy^2 + 8y^3 & (27x^2 + 18xy + 4y^2)
\end{array}
$$

The cube root of $27x^3$ is $3x$, the first term of the root. The trial divisor is $3(3x)^2 = 27x^2$.

The second term of the root is $\dfrac{54\,x^2y}{27\,x^2}, = 2\,y$. The complete divisor is

$$3(3\,x)^2 + 3(3\,x)(2\,y) + (2\,y)^2, = 27\,x^2 + 18\,xy + 4\,y^2,$$

which is multiplied by the second term of the root, giving

$$54\,x^2y + 36\,xy^2 + 8\,y^3.$$

21. The preceding method can be extended to find cube roots which are multinomials of any number of terms, as the method of finding square roots was extended. The work consists of repetitions of the following steps:

After one or more terms of the root have been found, obtain each succeeding term by dividing the first term of the remainder at that stage by three times the square of the first term of the root.

Find the next remainder by subtracting from the last remainder the expression $(3\,a^2 + 3\,ab + b^2)\,b$, wherein a stands for the part of the root already found, and b for the term last found.

22. The given multinomial should be arranged to powers of a letter of arrangement.

Ex.

```
27   27 x+90 x²−55 x³+90 x⁴−27 x⁵+27 x⁶ | 3−x+3 x²
27                                      |────────────
     ─────                               | 3(3)²=27
     −27 x
     −27 x+ 9 x²−   x³                   | 3(3)²+3(3)(−x)+(−x)²=27−9 x+x²
     ──────────────────                  | 3(3−x)²+3(3−x)(3 x²)+(3 x²)²=
          81 x²−54 x³                    | 27−18 x+30 x²−9 x³+9 x⁴
          81 x²−54 x³+90 x⁴−27 x⁵+27 x⁶  |
```

EXERCISES III.

Find the cube roots of following:

1. $a^3 - 6\,a^2 + 12\,a - 8.$ **2.** $a^3 + 6\,a^2b + 12\,ab^2 + 8\,b^3.$

3. $8\,a^3 + 36\,a^2b + 54\,ab^2 + 9\,b^3.$ **4.** $m^6 + 3\,m^4n^3 + 3\,m^2n^6 + n^9.$

5. $27\,m^3n^3 + 54\,m^2n^2p + 36\,mnp^2 + 8\,p^3.$

6. $64 a^3c^3 - 144 a^2c^2b + 108 acb^2 - 27 b^3$.

 7. $27 a^6b^3 + 27 a^4b^2c^2 + 9 a^2bc^4 + c^6$.

 8. $c^6 + 6 c^4 + 6 c^2 - 3 c^5 - 7 c^3 - 3 c + 1$.

 9. $x^6 + 3 x^5 - 5 x^3 + 3 x - 1$.

 10. $x^6 + 3 x^5 + 9 x^4 + 13 x^3 + 18 x^2 + 12 x + 8$.

 11. $m^6 - 3 m^5 + 9 m^4 - 13 m^3 + 18 m^2 - 12 m + 8$.

 12. $x^6 + 6 x^4y^2 + 6 x^2y^4 - 3 x^5y - 7 x^3y^3 - 3 xy^5 + y^6$.

ROOTS OF ARITHMETICAL NUMBERS.

Square Roots.

23. Since the squares of the numbers 1, 2, 3, ·, 9, 10, are 1, 4, 9, ·, 81, 100, respectively, the square root of an integer of *one or two* digits is a number of *one* digit.

Since the squares of the numbers 10, 11, ·, 100, are 100, 121, ···, 10000, the square root of an integer of *three or four* digits is a number of *two* digits; and so on.

Therefore, *to find the number of digits in the square root of a given integer, we first mark off the digits from right to left in groups of two. The number of digits in the square root will be equal to the number of groups, counting any one digit remaining on the left as a group.*

24. The method of finding square roots of numbers is then derived from the identity

$$(a + b)^2 = a^2 + (2a + b)b, \qquad (1)$$

wherein a denotes *tens* and b denotes *units*, if the square root is a number of two digits.

25. Ex. 1. Find the square root of 1296.

We see that the root is a number of *two* digits, since the given number divides into *two* groups. The digit in the *tens'* place is 3, the square root of 9, the square next less than 12. Therefore, in the identity (1), a denotes 3 *tens*, or 30.

154   ALGEBRA.   [Ch. XII

The work then proceeds as follows:

$$\begin{array}{r|l}
 & a+b \\
12'\,96 & 30+6=36 \\
9\,00 & 2a=60, \quad \text{trial divisor} \hfill (1) \\ \hline
3\,96 & (2ab+b^2) \div 2a = 396 \div 60 = 6+ \hfill (2) \\
3\,96 & = (2a+b) \times b = (60+6) \times 6 \hfill (3)
\end{array}$$

The first remainder, 396, is equal to $2ab + b^2$, and cannot be separated into the sum of two terms, one of which is $2ab$. We cannot, therefore, determine b by dividing $2ab$ by $2a$, as in finding square roots of algebraic expressions.

Consequently step (2) *suggests* the value of b, but does not definitely determine it. As a rule, we take the integral part of the quotient, 6 in the above example, and test that value by step (3).

This method may be extended to find roots which contain any number of digits. At any stage of the work a stands for the part of the root already found, and b for the digit to be found.

Ex. 2. Find the square root of 51529.

The root is a number of *three* digits, since the given number divides into *three* groups. The digit in the *hundreds'* place is 2, the square root of 4, the square next less than 5. Therefore in the identity (1), a denotes 2 *hundreds*, or 200, in the first stage of the work.

The work then proceeds as follows:

$$\begin{array}{r|l}
5'\,15'\,29 & 200+20+7=227 \\
4\,00\,00 & 2a=400, \quad \text{trial divisor} \hfill (1) \\ \hline
1\,15\,29 & (2ab+b^2) \div 2a = 11529 \div 400 = 20+ \hfill (2) \\
84\,00 & = (2a+b)b = (400+20) \times 20 \hfill (3) \\ \hline
31\,29 & (2ab+b^2) \div 2a = 3129 \div 440 = 7+ \hfill (4) \\
31\,29 & = (2a+b)b = (440+7) \times 7 \hfill (5)
\end{array}$$

In the second stage of the work, a stands for the part of the root already found, 220, and b for the next figure of the root. In practice the work may be arranged more compactly, omitting unnecessary ciphers, and in each remainder writing only the next group of figures. Thus:

$$\begin{array}{r|l}
5'\ 15'\ 29 & 227 \\
4 & \\
\hline
1\ 15 & 11 \div 4 = 2 + \qquad (2) \\
84 & 42 \\
\hline
31\ 29 & 312 \div 44 = 7 + \qquad (4) \\
31\ 29 & 447
\end{array}$$

Observe that the trial divisor at any stage is twice the part of the root already found, as in (2) and (4).

26. The abbreviated work in the last example illustrates the following method:

After one or more figures of the root have been found, obtain the next figure of the root by dividing the remainder at that stage (omitting the last figure) by the trial divisor at that stage.

See lines (2) and (4).

Annex this quotient to the part of the root already found, and also to the trial divisor to form the complete divisor.

Find the next remainder by subtracting from the last remainder the product of the complete divisor and the figure of the root last found.

27. Since the number of decimal places in the square of a decimal fraction is twice the number of decimal places in the fraction, the number of decimal places in the square root of a decimal fraction is one-half the number of decimal places in the fraction.

Consequently, in finding the square root of a decimal fraction, the decimal places are divided into groups of two from the decimal point to the right, and the integral places from the decimal point to the left as before.

Ex.

14'46.28'09	38.03
9	
5 46	
5 44	68
2.28 09	
2.28 09	76.03

In finding the second figure of the root, we have $\frac{54}{6} = 9$; but $69 \times 9 = 621$, which is greater than 546, from which it is to be subtracted. Hence we take the next less figure 8.

EXERCISES IV.

Find the square roots of the following:

1. 961.   2. 1681.   3. 1849.   4. 2704.
5. 4761. 6. 6724.   7. 9025.   8. 9801.
9. 39.69. 10. 54.76. 11. 10201. 12. 12321.
13. 18225. 14. 19881. 15. .005184. 16. .00000841.
17. .1225. 18. 10.24. 19. 96.04. 20. 219.04.

Cube Roots.

28. Since the cubes of the numbers 1, 2, 3, ·, 9, 10, are 1, 8, 27, ·, 729, 1000, respectively, the cube root of any integer of *one*, *two*, or *three* digits is a number of *one* digit. *The cube roots of such numbers can be found only by inspection.*

Since the cubes of 10, 11, ·, 100 are 1000, 1331, ·, 1000000, respectively, the cube root of any integer of *four*, *five*, or *six* digits is a number of *two* digits, and so on.

Therefore, to find the number of digits in the cube root of a given integer, we first mark off the digits from right to left in groups of *three*. The number of digits in the cube root will be equal to the number of groups, counting one or two digits remaining on the left as a group.

29. The method of finding cube roots of numbers is derived from the identity
$$(a + b)^3 = a^3 + (3a^2 + 3ab + b^2)b,$$
wherein a denotes *tens*, and b denotes *units*, if the cube root is a number of two digits.

Ex. Find the cube root of 59319.

The digits in the *tens'* place of the root is 3, the cube root of 27, the cube next less than 59. Therefore in identity (1) a denotes 3 *tens*, or 30. The work may be arranged as follows:

```
59'319 |  a + b
27 000 |  30 + 9
─────── 
32 319 | 3 a² = 3(30)² = 2700                                    (1)
       | (3 a²b + 3 ab² + b³) ÷ 3 a² = 32319 ÷ 2700 = 9 +         (2)
       |        3 a² = 3(30)²  = 2700
       |        3 ab = 3(30)9  =  810
       |         b²  =    9²   =   81
       |                       ─────
32 319 | = (3 a² + 3 ab + b²) × b = 3591 × 9                     (3)
```

As in finding square roots of numbers, step (2) *suggests* the value of b, but does not definitely determine it. If the value of b makes $(3a^2 + 3ab + b^2) \times b$ greater than the number from which it is to be subtracted, we must try the next less number.

In practice the work may be arranged more compactly, omitting unnecessary ciphers, and in each remainder writing only the next group of figures; thus

```
59' 319 | 39
 27     |
─────── 
32 319  | 2700                                                   (1)
        |  810                                                   (2)
        |   81                                                   (3)
        | ────
32 319  | 3591
```

30. The preceding method may be extended to find roots that contain any number of digits.

At any stage of the work a stands for the part of the root already found, and b for the digit to be found.

The method consists of repetitions of the following steps:

The trial divisor at any stage is three times the square of the part of the root already found; as 27 in the preceding example.

After one or more figures of the root have been found, obtain the next figure of the root by dividing the remainder at that stage (omitting the last two figures) by the trial divisor. In the last example, $9 + = 323 \div 27$.

Annex this quotient to the part of the root already found.

To get complete divisor, add to the trial divisor (with two ciphers annexed) three times the product of the part of the root already found (with one cipher annexed) by the figure of the root just found, and also the square of the figure of the root just found.

Find the next remainder by subtracting from the last remainder the product of the complete divisor and the figure of the root last found.

31. Evidently, in finding the cube root of a decimal fraction the decimal places are divided into groups of *three* figures from the decimal point to the right, and the integral places from the decimal point to the left as before.

Ex.

11'089.567	22.3
8	1200
3 089	120
	4
2 648	1324
441.567	1452.00
	19.80
	.09
441.567	1471.89

EXERCISES V.

Find cube roots of the following:

1. 1331. 2. 4913. 3. 9261. 4. 13824.
5. 2744. 6. 24389. 7. 59319. 8. 74088.
9. 103823. 10. 175616. 11. 15.625. 12. 704.969.

13. 39.304.   **14.** 314.432.   **15.** .456533.   **16.** .002197.
17. .006859.   **18.** 2146.689.   **19.** 3442951.   **20.** 4.492125.

Approximate Values of Irrational Roots.

32. Ex. 1. Find an approximate value of $\sqrt{2}$ correct to three decimal places. The work proceeds as follows:

```
2.00'00'00'00 | 1.4142
1             |
――――          | 2
1 00          |
  96          | 24
――――          |
 4 00         |
 2 81         | 281
――――          |
 1 19 00      |
 1 12 96      | 2824
――――          |
    6 04 00   | 2828
```

The work is simplified by neglecting the decimal point, writing it only in the result. It is necessary to find the root to four decimal places in order to determine whether to take the figure found in the third place or the next greater figure.

Ex. 2. Find an approximate value of $\sqrt{\tfrac{1}{2}}$ correct to three decimal places.

We have $\sqrt{\tfrac{1}{2}} = \sqrt{\tfrac{2}{4}} = \tfrac{\sqrt{2}}{\sqrt{4}} = \tfrac{1}{2}\sqrt{2}$, and $\sqrt{2} = 1.4142 + \cdots$

Therefore $\sqrt{\tfrac{1}{2}} = .707$, correct to three places of decimals.

EXERCISES VI.

Find to four decimal places the square roots of:

1. $\tfrac{3}{4}$.   **2.** $\tfrac{2}{3}$.   **3.** 21.   **4.** 2.5.
5. .075.   **6.** 6857.   **7.** 59.307.   **8.** .0017.

Find to three decimal places the cube roots of:

9. $\tfrac{3}{8}$.   **10.** $\tfrac{2}{3}$.   **11.** 2.   **12.** 2.5.
13. .17.   **14.** 959.   **15.** 1520.   **16.** .0029.

CHAPTER XIII.

QUADRATIC EQUATIONS.

1. A **Quadratic Equation** is an equation of the second degree in the unknown number or numbers.

E.g., $x^2 = 25$, $x^2 - 5x + 6 = 0$, $x^2 + 2xy = 7$.

A **Complete Quadratic Equation**, in one unknown number, is one which contains a term (or terms) in x^2, a term (or terms) in x, and a term (or terms) free from x, as $x^2 - 2ax + b = cx - d$.

A **Pure Quadratic Equation** is an incomplete quadratic equation which has no term in x, as $x^2 - 9 = 0$.

Pure Quadratic Equations.

2. Ex. 1. Solve the equation $6x^2 - 7 = 3x^2 + 5$.

Transferring $3x^2$ to the first member, and 7 to the second member,
$$6x^2 - 3x^2 = 5 + 7,$$
or
$$3x^2 = 12.$$
Dividing by 3, $\qquad x^2 = 4.$

The value of x is a number whose square is 4. But
$$2^2 = 4, \text{ and } (-2)^2 = 4.$$
Therefore $\qquad x = \pm 2.$

3. This example illustrates the following principle, which is proved in School Algebra, Ch. XXI.:

The positive square root of the first member of an equation may be equated in turn to the positive and to the negative square root of the second member.

Ex. 2. Solve the equation $ax^2 - a^3 = bx^2 - a^2b$.

Transferring terms, $\quad ax^2 - bx^2 = a^3 - a^2b$.

Factoring, $\quad (a-b)x^2 = a^2(a-b)$.

Dividing by $a-b$, $\quad x^2 = a^2$.

Equating square roots, $\quad x = \pm a$.

EXERCISES I.

Solve each of the following equations:

1. $x^2 = 25$.
2. $x^2 - 1 = 35$.
3. $2x^2 = x^2 + 49$.
4. $4x^2 - 18 = 2x^2$.
5. $5x^2 - 45 = 0$.
6. $x^2 - p^2 = 3p^2$.
7. $4x^2 = x^2 + 12a^2$.
8. $2x^2 - 1 = x^2 + 3$.
9. $4x^2 - 18 = x^2 + 9$.
10. $5x^2 - 1 = 3x^2 + 7$.
11. $3x^2 + 4 = 2x^2 + 5$.
12. $3(x^2 - 6) = 2(x^2 + 9)$.
13. $a^2(x^2 + 2) = 2a^2 + 4$.
14. $2(x^2 - x - 2) = x(x-2)$.

15. The product of three times and four times a certain number is 48. What is the number?

16. If 49 square feet is added to the area of a square floor, the area will be doubled. What is the area?

17. The product of the third and sixth parts of a certain number is 2. What is the number?

18. A woman has a square piece of oilcloth. If she takes a yard from its length, and adds it to its breadth, it just covers a room whose area is 24 square yards. What is the size of the square piece?

Solution by Factoring.

4. The principle on which the solution of an equation by factoring depends was proved in Ch. VI., Art. 24.

Ex. 2. Solve the equation
$$\frac{x}{x-7} + \frac{x+7}{4} = \frac{17}{2}.$$
Clearing of fractions,
$$4x + x^2 - 49 = 34x - 238.$$
Transferring and uniting terms,
$$x^2 - 30x + 189 = 0.$$
Factoring, $\quad (x-9)(x-21) = 0.$

Equating factors to 0,
$$x - 9 = 0, \text{ whence } x = 9;$$
$$x - 21 = 0, \text{ whence } x = 21.$$

EXERCISES II.

Solve each of the following equations:

1. $(x+1)(x-1) = 0.$
2. $(x - \frac{1}{5})(x - \frac{1}{7}) = 0.$
3. $x^2 - 64 = 0.$
4. $x^2 - \frac{1}{16} = 0.$
5. $x^2 + 3x - 4 = 0.$
6. $x^2 - 5x - 14 = 0.$
7. $x^2 - 4x - 21 = 0.$
8. $x^2 - 5x = 6x - 30.$
9. $x^2 - 10 + 2x = 11 \quad 2x.$
10. $x^2 + 12x - 4 = 5x - 16.$
11. $\dfrac{x}{3} + \dfrac{9}{x} = -4.$
12. $\dfrac{x}{4} - \dfrac{9}{x} = 4.$
13. $\dfrac{6}{x} + x = \dfrac{3x+4}{2}.$
14. $\dfrac{x+1}{x-2} = \dfrac{2x-3}{3x-10}.$
15. $\dfrac{3x+2}{2x+3} = \dfrac{5x-1}{3x+1}.$
16. $\dfrac{x-6}{x+30} = \dfrac{3}{x-6}.$
17. $\dfrac{4}{x-2} + \dfrac{6}{x-5} = 1.$
18. $\dfrac{3}{1+x} + \dfrac{3}{1-x} = 8.$
19. $\dfrac{x-1}{x} - \dfrac{3x}{x-1} = 2.$
20. $\dfrac{x+4}{x-4} + \dfrac{x-4}{x+4} = \dfrac{10}{3}.$

Solution by Completing the Square.

5. From the identity $(x+a)^2 = x^2 + 2ax + a^2$, we see that an expression like $x^2 + 2ax$ can be completed to the square of $x + a$ by adding a^2, *i.e.*, *the square of half the coefficient of x*.

Ex. Complete $x^2 + 6x$ to the square of a binomial. Adding $(\frac{6}{2})^2 = 3^2 = 9$, we have $x^2 + 6x + 9, = (x+3)^2$.

6. The following examples illustrate the solution of a quadratic equation by the method called *Completing the Square*.

Ex. 1. Solve the equation $x^2 - 5x + 6 = 0$.

Transferring 6, $\qquad x^2 - 5x = -6$.

To complete the square in the first member, we add $(-\frac{5}{2})^2$, $= \frac{25}{4}$, to this member, and therefore also to the second. We then have
$$x^2 - 5x + \tfrac{25}{4} = \tfrac{25}{4} - 6,$$
or $\qquad (x - \tfrac{5}{2})^2 = \tfrac{1}{4}$.

Equating square roots, $x - \tfrac{5}{2} = \pm \tfrac{1}{2}$, by Art. 3.

Whence $\qquad x = \tfrac{5}{2} \pm \tfrac{1}{2}$.

Therefore the required roots are $\tfrac{5}{2} + \tfrac{1}{2}, = 3$, and $\tfrac{5}{2} - \tfrac{1}{2}, = 2$.

Ex. 2. Solve the equation $7x^2 - 3x = 4$.

We wish the coefficient of x^2 to be unity.

Dividing by 7, $\qquad x^2 - \tfrac{3}{7}x = \tfrac{4}{7}$.

Adding $(-\tfrac{3}{2\times 7})^2, = \tfrac{9}{196}$, $\quad x^2 - \tfrac{3}{7}x + \tfrac{9}{196} = \tfrac{4}{7} + \tfrac{9}{196}$,

or, $\qquad (x - \tfrac{3}{14})^2 = \tfrac{121}{196}$.

Equating square roots, $\qquad x - \tfrac{3}{14} = \pm \tfrac{11}{14}$.

Whence $\qquad x = \tfrac{3}{14} \pm \tfrac{11}{14}$.

Therefore the required roots are

$$\tfrac{3}{14} + \tfrac{11}{14}, = 1, \text{ and } \tfrac{3}{14} - \tfrac{11}{14}, = -\tfrac{4}{7}.$$

7. The preceding examples illustrate the following method:

Bring the terms in x and x^2 to the first member, and the terms free from x to the second member, uniting like terms.

If the resulting coefficient of x^2 be not $+1$, divide both members by this coefficient.

Complete the square by adding to both members the square of half the coefficient of x.

Equate the positive square root of the first member to the positive and negative square roots of the second member.

Solve the resulting equations.

EXERCISES III.

Solve each of the following equations:

1. $x^2 + 2x = 3$.
2. $x^2 + 6x = 7$.
3. $x^2 + 8x = 9$.
4. $x^2 + 16x = 17$.
5. $x^2 - 4x = -3$.
6. $x^2 + x = 20$.
7. $x^2 - 7x + 10 = 0$.
8. $x^2 - 2x + 4 = 7$.
9. $x^2 + 2x - 6 = 2$.
10. $2x^2 - 7x + 3 = 0$.
11. $-3x^2 + x + 2 = 0$.
12. $5x^2 + 4x - 1 = 0$.
13. $6x^2 - 13x + 6 = 0$.
14. $-4x^2 + 4mx = -3m^2$.
15. $6x^2 - px - p^2 = 0$.
16. $4x^2 + 3lx - l^2 = 0$.
17. $\dfrac{x-1}{2} + \dfrac{1}{x} = 1$.
18. $\dfrac{x}{x+8} = \dfrac{3}{x-20}$.
19. $\dfrac{5+3x}{5+2x} = \dfrac{7+x}{7-x}$.
20. $\dfrac{x}{a+1} + \dfrac{1-a}{2a-x} = 0$.
21. $\dfrac{x}{7} + \dfrac{21}{x+5} = 3\tfrac{2}{7}$.
22. $\dfrac{4x}{x+3} - \dfrac{x-3}{2x+5} = 2$.

Problems.

8. Pr. The sum of two numbers is 15 and their product is 56. What are the numbers.

Let x stand for one of the numbers; then, by the first condition, $15 - x$ stands for the other number. By the second condition

$$x(15 - x) = 56; \text{ whence } x = 7, \text{ and } 8.$$

Therefore $x = 7$, one of the numbers, and $15 - x = 8$, the other number. Observe that if we take $x = 8$, then $15 - x = 7$.

That is, the two required numbers are the two roots of the quadratic equation.

9. When the solution of a problem leads to a quadratic equation, it is necessary to determine whether either or both of the roots of the equation satisfy the conditions expressed and implied in the problem.

It will, as a rule, be an easy matter to decide whether either or both of the results are admissible.

10. Pr. 1. Two men start at the same time to go from A to B, a distance of 36 miles. One goes 3 miles more an hour than the other, and arrives at B 1 hour earlier. At what rate does each man travel?

Let x stand for the number of miles the slower goes in 1 hour; then, $x+3$ stands for the number of miles the faster goes in 1 hour.

To go from A to B, it takes the slower $\dfrac{36}{x}$ of an hour, and the faster $\dfrac{36}{x+3}$ of an hour.

Therefore, by the condition of the problem,
$$\frac{36}{x} - \frac{36}{x+3} = 1.$$

Clearing of fractions, $36x + 108 - 36x = x^2 + 3x.$

Cancelling $36x$ and $-36x$, and interchanging members,
$$x^2 + 3x = 108.$$

Adding $(\tfrac{3}{2})^2, = \tfrac{9}{4},\ x^2 + 3x + \tfrac{9}{4} = 108 + \tfrac{9}{4},$

or $\qquad\qquad (x + \tfrac{3}{2})^2 = \tfrac{441}{4}.$

Equating square roots, $x + \tfrac{3}{2} = \pm \tfrac{21}{2}.$

Whence $\qquad x = -\tfrac{3}{2} + \tfrac{21}{2} = 9,$

and $\qquad x = -\tfrac{3}{2} - \tfrac{21}{2} = -12.$

Evidently -12 is not admissible.

Pr. 2. In a company of 14 persons, men and women, the men spent $20 and the women $20. If each man spent $3 more than each woman, how many men and how many women were in the company?

Let x stand for the number of men.

Then $14 - x$ stands for the number of women.

Each man spent $\dfrac{20}{x}$ dollars, and each woman $\dfrac{20}{14-x}$ dollars.

By the condition of the problem,
$$\frac{20}{x} = \frac{20}{14-x} + 3.$$

Clearing of fractions, $\qquad 280 - 20x = 20x + 42x - 3x^2$

Transferring and uniting terms,
$$3x^2 - 82x = -280.$$

Dividing by 3, $\qquad x^2 - \tfrac{82}{3}x = -\tfrac{280}{3}.$

Adding $(-\tfrac{41}{3})^2 = \tfrac{1681}{9}$, $\quad x^2 - \tfrac{82}{3}x + \tfrac{1681}{9} = -\tfrac{280}{3} + \tfrac{1681}{9}$,

or $\qquad\qquad\qquad\qquad (x - \tfrac{41}{3})^2 = \tfrac{841}{9}.$

Equating square roots, $\qquad x - \tfrac{41}{3} = \pm \tfrac{29}{3}.$

Whence $\qquad\qquad\qquad x = \tfrac{41}{3} + \tfrac{29}{3} = \tfrac{70}{3},$

and $\qquad\qquad\qquad\quad x = \tfrac{41}{3} - \tfrac{29}{3} = 4.$

Since the number of men must be an integer, $\tfrac{70}{3}$ is not admissible.

EXERCISES IV.

1. Divide 80 into two such parts that the difference of their squares is 1440.

2. One number is 10 greater than another. The sum of their squares is 1850. What are the numbers?

3. The factors of a product are 37 and 23. By what number must each factor be diminished in order that the product may may be diminished by 500?

4. The factors of a product are 73 and 31. By what number must the first factor be diminished and the second increased in order that the product may increased by 360?

5. Two men can do a piece of work in 18 days. How long will it take each to do it alone, if one requires 15 days more than the other?

QUADRATIC EQUATIONS.

6. A tank is supplied by 2 pipes. It will be filled in 6 hours if both pipes are opened. Running alone, 1 pipe requires 16 hours more than the other. How long will it take each pipe to fill the tank?

7. A woman brought eggs to market, and sold them for $3. Had she taken 10 fewer eggs, but sold them for the same amount, she would have received 1 penny more per egg. How many eggs had she?

8. A number of persons paid $24 for lodging at an inn. If the number of persons had been two less, and each person had paid 50 cents more, the whole bill would have been $1 more? How many persons were there?

9. From a thread whose length is equal to the perimeter of a square 36 inches are cut off, and the remainder is equal in length to the perimeter of another square whose area is $\frac{1}{9}$ of that of the first. What is the length of the thread?

10. A manufacturer paid $640 to 36 employees, men and women. Each man received as many dollars as there were women, and each woman as many dollars as there were men. How many men, and how many women were there?

11. A farmer sold a horse for $100. He lost $\frac{1}{4}$ as much per cent on its original cost as the horse cost dollars. What was the cost of the horse?

12. The diagonal of a rectangle, whose breadth is 119 metres shorter than its length, is 221 metres. What is the length and breadth of the rectangle?

13. An express train leaves New York for Philadelphia every morning at 6 o'clock. At the same time a freight train leaves Philadelphia for New York. When they meet, the express train has made $29\frac{2}{3}$ miles more than the freight train. How far is it from Philadelphia to New York, if the express train arrives 1 hour, and the freight train 4 hours, after they meet?

CHAPTER XIV.

SIMULTANEOUS QUADRATIC EQUATIONS.

1. Elimination by Substitution. — When one equation of a system of two equations is of the first degree, the solution can be obtained by the method of substitution.

Ex. Solve the system $y + 2x = 5$,   (1)

$\qquad\qquad\qquad\qquad x^2 - y^2 = -8.$ (2).

Solving (1) for y, $\qquad y = 5 - 2x.$ (3)

Substituting $5 - 2x$ for y in (2),

$$x^2 - 25 + 20x - 4x^2 = -8.$$

From this equation we obtain $\qquad x = 1,$

$$x = 5\tfrac{2}{3}.$$

Substituting 1 for x in (3), $\qquad y = 3.$

Substituting $5\tfrac{2}{3}$ for x in (3), $\qquad y = -6\tfrac{1}{3}.$

It is proved in School Algebra, Ch. XXIV., that the above method is based upon equivalent equations.

Therefore the solutions of the given system are 1, 3; $5\tfrac{2}{3}$, $-6\tfrac{1}{3}$, the first number of each pair being the value of x, and the second the corresponding value of y.

Had we substituted 1 for x in (2), we should have obtained $y = \pm 3$.

But the solution 1, -3 does not satisfy equation (1).

Therefore, always substitute in the linear equation the value of the unknown number obtained by elimination.

2. Elimination by Addition and Subtraction. — This method can frequently be applied.

Ex. Solve the system
$$x^2 + 3y = 18. \quad (1)$$
$$2x^2 - 5y = 3. \quad (2)$$

We will first eliminate y.

Multiplying (1) by 5, $\quad 5x^2 + 15y = 90.\quad(3)$

Multiplying (2) by 3, $\quad 6x^2 - 15y = 9.\quad(4)$

Adding (3) and (4), $\quad 11x^2 = 99.$

Whence, $\quad x = 3,$ and $x = -3.$

Substituting 3 for x in (1), $\quad y = 3.$

Substituting -3 for x in (1), $\quad y = 3.$

The given system has the two solutions 3, 3; -3, 3.

Notice that this example could also have been solved by the method of substitution.

3. Symmetrical Equations. — A **Symmetrical Equation** is one which remains the same when the unknown numbers are interchanged.

A system of two symmetrical equations can be solved by first finding the values of $x + y$ and $x - y$.

Ex. Solve the system
$$x^2 + y^2 = 13, \quad (1)$$
$$xy = 6. \quad (2)$$

Multiplying (2) by 2, $\quad 2xy = 12.\quad(3)$

Adding (3) to (1), $\quad x^2 + 2xy + y^2 = 25.\quad(4)$

Subtracting (3) from (1), $x^2 - 2xy + y^2 = 1.\quad(5)$

Equating square roots of (4), $\quad x + y = \pm 5.\quad(6)$

Equating square roots of (5), $\quad x - y = \pm 1.\quad(7)$

It is proved in School Algebra, Ch. XXIV., that equations (6) and (7) are equivalent to

$$\begin{matrix}x+y=5,\\x-y=1,\end{matrix}\quad \begin{matrix}x+y=5,\\x-y=-1,\end{matrix}\quad \begin{matrix}x+y=-5,\\x-y=+1,\end{matrix}\quad \begin{matrix}x+y=-5,\\x-y=-1.\end{matrix}$$

The solutions of these four systems are respectively 3, 2; 2, 3; −2, −3; −3, −2.

The solutions of (6) and (7) should be obtained mentally, without writing the equivalent systems. Each sign of the second member of (6) should be taken in turn with each sign of the second member of (7).

EXERCISES I.

Solve the following equations:

1. $\begin{cases} x^2 + y^2 = 13, \\ x^2 - y^2 = 5. \end{cases}$
2. $\begin{cases} x^2 + y^2 = 25, \\ x^2 - y^2 = 7. \end{cases}$

3. $\begin{cases} x^2 + y^2 = 10, \\ 3x^2 - 5y^2 = 22. \end{cases}$
4. $\begin{cases} x^2 - y^2 = 16, \\ x^2 + 3y^2 = 5'' \end{cases}$

5. $\begin{cases} x^2 - 5y = 10, \\ x^2 + 3y = 34. \end{cases}$
6. $\begin{cases} x^2 + y^2 = 100, \\ 3y = 4x. \end{cases}$

7. $\begin{cases} x + y = 5, \\ xy = 6. \end{cases}$
8. $\begin{cases} x + y = 7, \\ xy = 12. \end{cases}$

9. $\begin{cases} x^2 + y^2 = 13, \\ x + y = 5. \end{cases}$
10. $\begin{cases} x - y = 7, \\ xy = 8. \end{cases}$

11. $\begin{cases} x - 2y = 2, \\ xy = 12. \end{cases}$
12. $\begin{cases} x^2 - xy = 153, \\ x + y = 1. \end{cases}$

13. $\begin{cases} \dfrac{1}{x} + \dfrac{1}{y} = 1, \\ x + y = 4. \end{cases}$
14. $\begin{cases} \dfrac{1}{x} + \dfrac{1}{y} = 2, \\ x + y = 2. \end{cases}$

15. $\begin{cases} x^2 + y^2 = 65, \\ xy = 28. \end{cases}$
16. $\begin{cases} x^2 + y^2 = 8, \\ \dfrac{1}{x^2} + \dfrac{1}{y^2} = \dfrac{1}{2}. \end{cases}$

17. $\begin{cases} x^2 + xy + y^2 = 19, \\ x + y = 5. \end{cases}$
18. $\begin{cases} x^2 - xy + y^2 = 7, \\ x + y = 4. \end{cases}$

19. $\begin{cases} x^2 - xy = 153, \\ x + y = 1. \end{cases}$ **20.** $\begin{cases} x^2 - y^2 = 33, \\ x - y = 3. \end{cases}$

21. $\begin{cases} x + y = c, \\ xy = \dfrac{c^2}{4} - d^2. \end{cases}$ **22.** $\begin{cases} \dfrac{1}{x} + \dfrac{1}{y} = \dfrac{1}{2}, \\ \dfrac{2}{xy} = \dfrac{1}{9}. \end{cases}$

Problems.

4. Pr. The front wheel of a carriage makes 6 more revolutions that the hind wheel in travelling 360 feet. But if the circumference of each wheel were 3 feet greater, the front wheel would make only 4 revolutions more than the hind wheel in travelling the same distance as before. What are the circumferences of the two wheels?

Let x stand for the number of feet in the circumference of front wheel, and y for the number of feet in the circumference of hind wheel. Then in travelling 360 feet the front wheel makes $\dfrac{360}{x}$ revolutions, and the hind wheel makes $\dfrac{360}{y}$ revolutions.

By the first condition, $\dfrac{360}{x} = \dfrac{360}{y} + 6.$ (1)

If 3 feet were added to the circumference of each wheel, the front wheel would make $\dfrac{360}{x+3}$ revolutions, and the hind wheel $\dfrac{360}{y+3}$ revolutions.

By the second condition, $\dfrac{360}{x+3} = \dfrac{360}{y+3} + 4.$ (2)

Whence $x = 12$, the circumference of the front wheel, and $y = 15$, the circumference of the hind wheel.

EXERCISES II.

1. The sum of the squares of two numbers is 41. The difference of their squares is 9. What are the numbers?

2. The square of the sum of two numbers is 49. The sum of their squares is 25. What are the numbers?

3. A man has two square fields. If he plants them both in barley, he has 25 square rods of barley. If he plants one in oats and the other in barley, he has 7 square rods more of barley than of oats. What is the size of each field?

4. A man has a square closet adjoining a square room. The area of the room and closet is 29 square yards, and five times a side of the closet is equal to twice a side of the room. What is the size of each?

5. The distance around a rectangular field is 36 rods. The area is 80 square rods. What are its dimensions?

6. The sum of the squares of two numbers is 1690. If each number were increased by 2, the sum of the squares would equal 1930. What are the numbers?

7. Divide 12 into two parts so that the sum of their squares is 80.

8. A man paid $3 for a number of lemons. If he had bought 10 more lemons for the same money, they would have cost him 1 cent apiece less. How many did he buy, and what was the cost of each lemon?

9. Two square surfaces contain together 100 square feet. Placed one above the other, their height is 14 feet. What are the dimensions of the surfaces?

10. I have a certain number of silver dollars. If I lay them on the table in the form of a square, it must have 10 on a side. If I form two squares, one contains 28 more dollars than the other. How many dollars are there in each side of the small square?

11. A man has fencing for a rectangular field of 1375 square rods. Some of it is burned, and he finds he can fence another field 6 rods less each way, the area of which is 931 square yards. What are the dimensions of the original field?

12. A number of persons paid $42 for lodging at an inn. If there had been 3 more persons, and each person had paid 50 cents more, the total payment would have been $60. How many persons were there, and what did each pay?

CHAPTER XV.

PROGRESSIONS.

1. A **Series** is a succession of numbers, each formed according to some definite law. The single numbers are called the **Terms** of the series.

E.g., in the series
$$1 + 3 + 5 + 7 + 9 + \cdots \tag{1}$$
each term after the first is formed by adding 2 to the preceding term.

In the series $\quad 1 + 2 + 4 + 8 + \cdots \tag{2}$

each term after the first is formed by multiplying the preceding term by 2.

ARITHMETICAL PROGRESSION.

2. An **Arithmetical Series**, or, as it is more commonly called, an **Arithmetical Progression** (A. P.), is a series in which each term, after the first, is formed by adding a constant number to the preceding term. See Art. 1, (1).

3. Evidently this definition is equivalent to the statement, that the difference between any two consecutive terms is constant.

E.g., in the series
$$1 + 3 + 5 + 7 + \cdots$$
we have $\quad 3 - 1 = 5 - 3 = 7 - 5 = \cdots$

For this reason the constant number of the first definition is called the **Common Difference** of the series.

4. Let a_1 (read *a sub-one*, or *a one*) stand for the first term of the series,

a_n (read *a-n*) for the *n*th (*any*) term of the series,

d for the common difference,

and S_n (read *S-n*) for the sum of *n* terms of the series.

5. The common difference may be either positive or negative. If d be *positive*, each term is greater than the preceding, and the series is called a *rising*, or an *increasing* progression.

E.g., $1 + 2 + 3 + 4 + \cdots$, wherein $d = 1$.

If d be negative, each term is less than the preceding, and the series is called a *falling*, or a *decreasing* progression.

E.g., $1 - 1 - 3 - 5 - \cdots$, wherein $d = -2$.

The *n*th Term of an Arithmetical Progression.

6. By the definition of an arithmetical progression,

$$a_1 = a_1,\ a_2 = a_1 + d,\ a_3 = a_2 + d = a_1 + 2d,\ \text{etc.}$$

The law expressed by the formulæ for these first three terms is evidently general, and since the coefficient of d in each is one less than the number of the corresponding term, we have

$$a_n = a_1 + (n-1)d. \tag{I.}$$

That is, to find the *n*th term of an arithmetical progression: *Multiply the common difference by $n - 1$, and add the product to the first term.*

7. Ex. 1. Find the 15th term of the progression,

$$1 + 3 + 5 + 7 + \cdots.$$

We have $a_1 = 1,\ d = 2,\ n = 15$;

therefore $a_{15} = 1 + (15 - 1)2 = 1 + 28 = 29.$

ARITHMETICAL PROGRESSION.

This formula may be used not only to find a_n, when a_1, d, and n are given, but also to find any one of the four numbers involved when the other three are given.

Ex. 2. If $a_5 = 3$ $(n = 5)$, and $a_1 = 1$, we have $3 = 1 + 4d$; whence $d = \frac{1}{2}$.

EXERCISES I.

1. Find the 10th term of the series 2, 4, 6, ⋯.
2. Find the 21st term of the series 6, 10, 14, ⋯.
3. Find the 16th term of the series 8, 5, 2, ⋯.
4. Find the 12th term of the series $\frac{1}{4}$, $\frac{3}{4}$, $\frac{5}{4}$, ⋯.
5. Find the 15th term of the series $\frac{2}{3}$, $-\frac{1}{3}$, $-\frac{5}{3}$, ⋯.
6. Given $a_1 = 7$, $a_n = 16$, $n = 4$. Find d.
7. Find the 12th term of a series whose 1st term is 6, and whose 4th term is 12.
8. Find the 15th term of a series whose 1st term is 5, and whose 3d term is -1.
9. Given $a_n = 29$, $a_1 = 2$, $d = 3$. Find n.
10. 99 is what term of the series 4, 9, 14, ⋯?
11. $-\frac{31}{4}$ is what term of the series $\frac{1}{4}$, $-\frac{1}{4}$, $-\frac{3}{4}$, ⋯?
12. -40 is what term of the series 8, 5, 2, ⋯?
13. Given $a_n = l$, $n = n$, $d = 2$. Find a_1.
14. What is the 1st term of the series whose 16th term is 51, and whose common difference is 3?
15. What is the 1st term of the series whose 55th term is -168, and whose common difference is -3?

The Sum of n Terms of an Arithmetical Progression.

8. The successive terms in an arithmetical progression, from the first to the nth inclusive, may be obtained either by repeated additions of the common difference beginning with the first term, or by repeated subtractions of the common difference beginning with the nth term. We may therefore express the sum of n terms in two equivalent ways:

$$S_n = a_1 + (a_1 + d) + (a_1 + 2d) + \cdots + (a_1 + \overline{n-2} \cdot d) + (a_1 + \overline{n-1} \cdot d),$$
$$S_n = a_n + (a_n - d) + (a_n - 2d) + \cdots + (a_n - \overline{n-2} \cdot d) + (a_n - \overline{n-1} \cdot d).$$

Whence, by addition,

$$2 S_n = (a_1 + a_n) + (a_1 + a_n) + \cdots + (a_1 + a_n) + (a_1 + a_n),$$

wherein there are n binomials $a_1 + a_n$.

Therefore, $2 S_n = n(a_1 + a_n)$, or $S_n = \dfrac{n}{2}(a_1 + a_n)$. (II.)

9. If the value of a_n, given in (I.), be substituted for a_1 in (II.), we obtain

$$S_n = \frac{n}{2}[2a_1 + (n-1)d]. \qquad \text{(III.)}$$

Formula (II.) is used when a_1, a_n, and n are given; and (III.) when a_1, d, and n are given.

10. Ex. 1. If $a_1 = 1$, $a_5 = 3$, then $S_5 = \tfrac{5}{2}(1+3) = 10$.

Ex. 2. If $a_1 = -4$, $d = 2$, $n = 12$,

then $S_{12} = \tfrac{12}{2}[2(-4) + 11 \times 2] = 84$.

Either (II.) or (III.) can be used to determine any one of the five elements a_1, a_n, d, n, S_n, when the three others involved in the formula are known.

Ex. 3. Given $a_1 = -3$, $d = 2$, $S_n = 12$, to find n.

From (III.), $12 = \dfrac{n}{2}[-6 + 2(n-1)]$,

or $n^2 - 4n = 12$; whence $n = 6$ and -2.

The result 6 gives the series $-3 - 1 + 1 + 3 + 5 + 7, = 12$.

Since the number of terms must be positive, the negative result, -2, is not admissible.

Arithmetical Means.

11. The **Arithmetical Mean** between two numbers is a third number, in value between the two, which forms with them an arithmetical progression.

E.g., 2 is an arithmetical mean between 1 and 3.

Let A stand for the arithmetical mean between a and b; then, by the definition of an arithmetical progression,

$$A - a = b - A,$$

whence
$$A = \frac{a+b}{2}$$

That is, the arithmetical mean between two numbers is half their sum.

12. Arithmetical Means between two numbers are numbers, in value between the two, which form with them an arithmetical progression.

E.g., 2, 3, and 4 are three arithmetical means between 1 and 5.

Ex. Insert four arithmetical means between -2 and 9.

We have $n = 6$, $a_1 = -2$, $a_6 = 9$.

From (I.), $9 = -2 + 5d$, whence $d = \tfrac{11}{5}$.

The required means are $\tfrac{1}{5}$, $1\tfrac{2}{5}$, $2\tfrac{3}{5}$, $3\tfrac{4}{5}$.

EXERCISES II.

1. Find the sum of 20 terms of 7, 10, 13, ⋯.

2. Find the sum of 15 terms of 10, 7, 4, ⋯.

3. Find the sum of 13 terms of $\tfrac{1}{4}$, $\tfrac{7}{12}$, $\tfrac{11}{12}$, ⋯.

4. Find the sum of 10 terms of 14, $13\tfrac{1}{2}$, 13, ⋯.

5. Given $a_n = 20$, $a_1 = 2$, $S_n = 77$. Find n.

6. How many terms of the series 4, 7, 10, ⋯ **must I take** to obtain a sum equal to 650?

7. How many terms of the series 5, 1, −3, ⋯ must I take to obtain a sum equal to −1824?

8. Find the sum of m terms of the series whose first term is l, and whose common difference is 2.

9. Insert 10 arithmetical means between 13 and 46.

10. Insert 8 arithmetical means between −8 and −35.

11. Insert 12 arithmetical means between $-\frac{1}{4}$ and $-\frac{27}{4}$.

12. What is the arithmetical mean of $a^2 + b^2$ and $a^2 - b^2$?

Problems.

13. Pr. Find the sum of all the numbers of three digits which are multiples of 7.

The numbers of three digits which are multiples of 7 are

$$7 \times 15,\ 7 \times 16,\ 7 \times 17,\ \cdot,\ 7 \times 142.$$

Their sum is $\quad 7(15 + 16 + \cdots + 142)$.

The series within the parentheses is an arithmetical progression, in which $a_1 = 15$, $d = 1$, $n = 128$, and $a_{128} = 142$.

Therefore $\quad\quad\quad S_{128} = 10048.$

The required sum is therefore $7 \times 10048, = 70336$.

EXERCISES III.

1. Find the sum of the first 50 natural numbers.

2. A gentleman employs a servant for 12 years. He promises him $150 the first year, and an increase of $10 each succeeding year. How much does the servant receive the last year, and what amount has he received for his 12 years' services?

3. A man is employed to bore a well 500 feet deep. On account of the ground he is to receive $3.24 for the first foot and 5 cents additional for each succeeding foot. How much does the man receive, and what does the last foot cost?

GEOMETRICAL PROGRESSION.

4. A clock strikes the hours from 1 to 12, and once at each half hour. How many times does it strike in 24 hours?

5. The sum of three numbers in arithmetical progression is 33. The product of the two extremes is 72. What are the numbers?

6. A body falling freely from a height passes over 16 feet in the first second and 32 feet more in each succeeding second than in the preceding one. How far will the body fall in 12 seconds?

GEOMETRICAL PROGRESSION.

14. A **Geometrical Series**, or, as it is more commonly called, a **Geometrical Progression** (G. P.), is a series in which each term after the first is formed by multiplying the preceding term by a constant number. See Art. 1, (2).

15. Evidently this definition is equivalent to the statement that the ratio of any term to the preceding is constant.

For this reason the constant multiplier of the first definition is called the **Ratio** of the progression.

16. Let a_1 stand for the first term of the series,

a_n for the nth (any) term,

r for the ratio,

and S_n for the sum of n terms.

17. The ratio may be either larger or smaller than 1; in the former case the progression is called a *rising* or *ascending* progression; in the latter a *falling* or *descending* progression.

E.g., $1 + \frac{3}{2} + \frac{9}{4} + \frac{27}{8} + \cdots$, in which $r = \frac{3}{2}$,

and $\frac{1}{2} - 1 + 2 - 4 + 8 \cdots$, in which $r = -2$

are *ascending* progressions; while

$1 + \frac{1}{2} + \frac{1}{4} + \frac{1}{8} + \cdots$, in which $r = \frac{1}{2}$,

and $1 - \frac{2}{3} + \frac{4}{9} - \frac{8}{27} + \cdots$ in which $r = -\frac{2}{3}$,

are *descending* progressions.

The nth Term of a Geometrical Progression.

18. By the definition of a geometrical progression

$$a_1 = a_1, \quad a_2 = a_1 r, \quad a_3 = a_2 r = a_1 r^2, \quad a_4 = a_3 r = a_1 r^3, \text{ etc.}$$

The law expressed by the relations for these first four terms is evidently general, and since the exponent of r in each is one less than the number of the corresponding term, we have

$$a_n = a_1 r^{n-1}. \qquad \text{(I.)}$$

That is, to find the nth term of a Geometrical Progression: *Raise the ratio to a power one less than the number of the term, and multiply the result by the first term.*

Ex. 1. If $a_1 = \frac{1}{2}, r = 3, n = 5$, then $a_5 = \frac{1}{2} \cdot 3^4 = \frac{81}{2}$.

This relation may also be used to find not only a_n, when a_1, r, and n are given, but also to find the value of any one of the four numbers when the other three are given.

Ex. 2. If $a_1 = 4, a_6 = \frac{1}{8}, n = 6$, then $\frac{1}{8} = 4 r^5$, whence $r = \frac{1}{2}$.

EXERCISES IV.

1. Find the 6th term of $2, 6, 18, \cdots$.
2. Find the 5th term of $3, 12, 48, \cdots$.
3. Find the 8th term of $\frac{1}{2}, -\frac{1}{4}, \frac{1}{8}, \cdots$.
4. Find the 7th term of $\dfrac{a}{3}, \dfrac{a^2}{9}, -\dfrac{a^3}{27}, \cdots$.
5. The third term is 369; the 5th term is 3321. What is the common ratio?

The Sum of a Geometrical Progression

19. We have $S_n = a_1 + a_1 r + a_1 r^2 + \cdots + a_1 r^{n-2} + a_1 r^{n-1}$, (1)

and $\quad rS_n = \quad a_1 r + a_1 r^2 + \cdots + a_1 r^{n-2} + a_1 r^{n-1} + a_1 r^n$ (2)

Consequently, subtracting (2) from (1),

$$S_n(1-r) = a_1 - a_1 r^n,$$

whence $\quad S_n = \dfrac{a_1(1-r^n)}{1-r} = \dfrac{a_1(r^n-1)}{r-1}. \qquad \text{(II.)}$

Substituting a_n for $a_1 r^{n-1}$ in (II.), we have

$$S_n = \frac{a_1 - a_n r}{1 - r} = \frac{a_n r - a_1}{r - 1}. \qquad \text{(III.)}$$

The first forms of (II.) and (III.) are to be used when $r < 1$, the second when $r > 1$.

20. Ex. 1. Given $a_1 = 3$, $r = 2$, $n = 6$, to find S_6.

From (II.), $\quad S_6 = \dfrac{3(2^6 - 1)}{2 - 1} = 189$.

Formulæ (II) and (III.) may be used not only to find S_n when a_1, r, and n, or a_1, a_n, and r are given, but also to find the value of any one of the four numbers when the other three are given.

Ex. 2. Given $S_n = -63\frac{1}{2}$, $a_1 = -\frac{1}{2}$, $a_n = -32$, to find r.

By (III.), $\quad -63\frac{1}{2} = \dfrac{-\frac{1}{2} + 32 r}{1 - r}$, whence $r = 2$.

Geometrical Means.

21. A **Geometrical Mean** between two numbers is a number, in value between the two, which forms with them a geometrical progression.

E.g., $+2$, or -2, is a geometrical mean between 1 and 4.
Let G be the geometrical mean between a and b.
Then by definition of a geometrical progression,

$$\frac{G}{a} = \frac{b}{G}; \text{ whence } \boldsymbol{G} = \pm \sqrt{(ab)}.$$

That is, the geometrical mean between two numbers is the square root of their product.

Ex. Find the geometrical mean between 1 and $\frac{4}{9}$. We have

$$G = \pm \sqrt{(1 \times \tfrac{4}{9})} = \pm \tfrac{2}{3}.$$

22. Geometrical means between two numbers are numbers, in value between the two, which form with them a geometrical progression. *E.g.*, 4 and 16 are two geometrical means between 1 and 64; and 2, 4, 8, 16, 32 are five geometrical means between 1 and 64.

Ex. Insert 5 geometrical means between 1 and 729.

We have $a_1 = 1, n = 7, a_n = 729$.

Therefore $729 = r^6$, or $r = \pm 3$.

The required means are:

$$\pm 3, 9, \pm 27, 81, \pm 243.$$

EXERCISES V.

1. Find the sum of 2, 6, 18, ·, to 7 terms.
2. Find the sum of 3, 9, 27, ·, to 8 terms.
3. Find the sum of 4, 12, 36, ·, to 6 terms.
4. Find the sum of 5, 1, $\frac{1}{5}$, ·, to 6 terms.
5. Find the geometrical mean of 50 and 648.
6. Find the geometrical mean of $\frac{1}{3}$ and $\frac{1}{12}$.
7. Find the geometrical mean of xy and $\frac{x}{y} + \frac{y}{x}$ 2.

Problems.

23. Pr. A farmer agrees to sell 12 sheep on the following terms: he is to receive 2 cents for the first sheep, 4 cents for the second, 8 cents for the third, and so on. How much does he receive for the twelfth sheep, how much for the 12 sheep, and what is the average price?

We have $a_1 = 2, n = 12, r = 2$.

Then $a_{12} = 2 \times 2^{11} = 2^{12} = 4096.$

And $S_{12} = \dfrac{2(2^{12} - 1)}{2 - 1} = 2 \times 4095 = 8190.$

That is, he receives 4096 cents, or $40.96, for the twelfth sheep, and 8190 cents, or $81.90, for the 12 sheep.

The average price is $\frac{81.90}{12}$, = $6.82½.

EXERCISES VI.

1. A contractor takes a piece of work on the condition that if he finishes 1 day ahead of time he shall receive $3; 2 days ahead of time $6; 3 days ahead of time $12, and so on. He finishes 8 days ahead of time. How much extra does he receive?

2. The population of a western town 6 years ago was 8000. It has been increasing at the rate of 5 per cent a year. What is its present population, neglecting decimals?

3. What would $2 amount to at compound interest for 6 years at 5 per cent?

4. A diver for pearls is to receive $3 for the first, $4.50 for the second, $6.75 for the third, and so on. He finds eight. How much money does he receive?

5. The geometrical mean of two numbers is 9; the arithmetical mean is 15. What are the numbers?